Anna Comacchio · Giuseppe Volpato · Arnaldo Camuffo

Automation in Automotive Industries

Springer

Berlin
Heidelberg
New York
Barcelona
HongKong
London
Milano
Paris
Singapore
Tokyo

Anna Comacchio
Giuseppe Volpato
Arnaldo Camuffo (Eds.)

Automation in Automotive Industries

Recent Developments

With 9 Figures

Springer

Professor Anna Comacchio
Professor Giuseppe Volpato
Professor Arnaldo Camuffo
Department of Business Economics and Management
University Ca´ Foscari
San Trovaso 1075
30123 Venice
ITALY

e-mail: acomac@unive.it
volpato@unive.it
camuffo@unive.it

ISBN 3-540-64018-5 Springer-Verlag Berlin Heidelberg New York

Cataloging-in-Publication Data applied for

Die Deutsche Bibliothek – CIP-Einheitsaufnahme

Automation in Automotive Industries: Recent Developments / Anna Comacchio; Giuseppe Volpato; Arnaldo Camuffo (Hrsg.). – Berlin; Heidelberg; New York; Barcelona; Hong Kong; London; Milano; Paris; Singapore; Tokyo: Springer, 1999
ISBN 3-540-64018-5

Typesetting: Camera-ready by authors
Cover-Design: MEDIO GmbH, Berlin
SPIN: 10662448   68/3020-5 4 3 2 1 0 - Printed on acid -free paper

Contents

1 Introduction

G. Volpato, A. Camuffo, A. Comacchio

1.1
The background

During recent years the dynamics of automotive industry and its supply chain has catalysed the attention and the research effort of a wide international group of scholars as: the International Motor Vehicle Program (IMVP) of Massachusetts Institute of Technology, the Permanent Study Group for the Automobile Industry and Its Employees (GERPISA) of Paris, and the International Car Distribution Programme (ICDP) of Solihull. This favoured the publication of relevant studies[1] and the growth of networks of academicians and practitioners interested in studying the patterns of industry evolution and in organising meetings to present and discuss issues of common interest.

In 1992 some members of these research projects decided to organize a first conference in Berlin dedicated to the main theme of automation and organization in the automobile industry. In 1993 a second conference took place in Tokyo, followed by a technical visit to a few automobile manufacturers and components suppliers plants (Toyota, Nissan, Mitsubishi, etc.).

After the two conferences, the colleagues of the network invited the Department of Business Economics and Management of the "Ca' Foscari" University of Venice to organize in Italy a new conference. Thus the Third Automation Conference "Manufacturing systems and organizational paradigm in automobile industry: international patterns of diffusion" was held in Venice in October 1995. This book collects a part of the proceedings of the conference which enjoyed the attending of a vast number of international scholars and practitioners and the visit to the greenfield Fiat Auto plant in Melfi (South Italy).

---

[1] Altshuler et al. 1984, Womack, Jones and Roos, 1990; Womack and Jones 1996, Kochan, MacDuffie and Lansbury 1996; Shimokawa, Jurgens and Fujimoto 1997, Freyssenet, Mair, Shimizu, Volpato 1998.

The papers and the case studies at the Italian Conference dealt with the relationships among automation, automobile production technology and plant organization. More specifically the main issues were:
- General Overview and Scenarios for Auto Manufacturing
- Patterns of Manufacturing and Automation Adoption
- Automation in Auto Parts Manufacturing
- Human Resource Management for Competitive Manufacturing
- Manufacturing Systems in Auto Industry: convergence or divergence?

We believe that the discussion brought forward by the three conferences and the further considerations raised by the publication of the proceedings have a manifold function. Firstly they can help to bridge research and practice in the auto industry through the exchange of ideas and experiences between the academic world and the professionals of the automobile industry involved in a huge reorganization process. Secondly they are an important opportunity to exchange points of view and present the last results achieved in each international research project, maintaining a fruitful collaboration among international scholars of the automobile industry. Finally, the subsequent research presentation and the publication of the proceedings represent a "state of the art" in the most advanced international research dealing with relationships among automation, automobile production technology and plant organization.

1.2
Aim of the book

In reference to the original work of international researchers from Europe, US and Japan, the book focuses on understanding the most recent change patterns in automation, manufacturing systems, organization and management of human resources of the world automobile firms.

More specifically one important driver of the technological and organizational change is the diffusion of the "lean management system". Some researchers discuss whether or not it is the new production paradigm, but the focus of the book is rather different and it adopts a dynamic perspective. The patterns of adoption are analyzed at industry, firm and plant level and the convergent or divergent experiences are discussed.

Another relevant driver of technological and organizational change is the increasing competitive pressure that auto makers began to face in the middle of the Nineties and are facing in the last few years. Invested by the globalization process, changing product and labor market conditions as well as the economic crisis at national level, firms are striving to defend their competitive position.

In brief, even if the automation issue is one of the main focuses of this book, regarding specific technological solutions, we refer the readers to the in depth analysis of the proceedings of previous conferences. This book deals with adjustment processes underlying significant experiences. Specifically, the main aim of the book is to understand the new step of the evolutionary process which involves the manufacturing system of the automobile supply chain (automobile manufacturers and parts manufacturing firms). The drivers of this evolution are analyzed, specifically focusing on the continuos interaction among changing competitive context, new technological patterns and emerging human resource management and organizational issues.

1.3
Contributions

The book is organized in seven chapters. The papers included in this book, are few pieces of work selected from several significant contributions presented at the conference of Venice. In revising the papers for this book the authors took into account ideas and discussions emerged during the conference. For this reason in presenting each chapter, we wish to thank all practitioners and academicians whose participation at the conference was so fruitful.

The first five chapters of the book deal with the dynamic of automation in the automobile supply chain (Fine). This dynamic is related to the role of automation in competitive performance. In this perspective the authors consider the relationship among firm automation strategy and competitive environment (Volpato), competitive strategy and industrial form of the firm (Freyssenet), organizational and individual experiences (Ellegard), and supplier automation strategy (Amikura). The authors analyze the role of these factors on competitive performance of the firm and the convergence and divergence of the experiences.

In the first chapter the introductory essay of Fine draws the attention on the dynamic of the industries and on the importance to manage the supply chain coherently with the industry clockspeed.

According with this perspective Volpato, in the next chapter, analyses the recent dynamic of the world automotive supply chain. Volpato gives a wide overview of the most recent competitive evolution at industry level (for instance globalization, modularization of production, quality). From this perspective he analyzes the possible impact of these changes on automation strategies of auto makers and their suppliers.

In the third chapter, M. Freyssenet relies on the results of the international research program of GERPISA, carried out from 1993 to 1996, to discuss the diversities of strategies and automation templates from an evolutionary perspective. He highlights the role of history, especially in firm industrial models but also but also mismatches among industrial model, automation form and the chosen competitive strategy, to explain the divergence of firm experiences.

The paper of Ellegard deals with the individual learning process triggered by technological innovation. Ellegard studies the automation development of the Volvo Torslanda bodyshop from 1970 to 1990, considering how workers and managers previous experiences influence their approach and reaction to automation.

Amikura set a theoretical framework for a research project on automation of auto suppliers in Japan. The project is at a pilot phase. The paper discusses early results and sets the direction for future research. Based on the premise that the "tiered" structure of Japanese subcontracting is a major source of Japanese automakers' competitive advantage, Amikura aims to explore the congruence of assembly automation of assemblers and part suppliers related to achieving the apparent trade-off between flexibility and efficiency.

As mentioned above, the approach adopted by the book is systemic and evolutionary. Improvements at automation level this way may not results in a linear predictable way. Competitive performance is related to the interaction of several factors along the added value chain and to the institutional and economic environment outside the firm.

For instance, as product cycle times reduce and time-to-market of new products becomes shorter, flexibility at production level probably needs to match the solutions adopted in previous phases, namely the product development process. Starting from this problem, Jurgens' chapter considers issues to improve production performance by anticipating manufacturing problems during the product development process. Jurgens deals with the main approaches and organizational solutions, drawn from his research in German and U.S. car companies as part of an internationally comparative project on new product and process development networks.

The next paper by Camuffo and Comacchio adopts the same approach, considering the role of human resource management on competitive performance of the firm. The authors argue that, although automation represents a long term inarrestable trend, the search of competitiveness is tending towards continuous, but more cautious, investment in flexible automation and adoption of lean management practices at an organizational level. The authors sustain that these processes are firm specific. They argue that the firms simply do not imitate, but rather they enact a specific, contingent creative combination of firm policies and new techniques which result in a variety of technological and organizational models. Moreover, as far as the adoption process is concerned, both institutional and competitive context are important. Evidence from the European auto industry show how the "lean" concept tends to evolve as it is implemented in different contexts.

The problem of diffusion of lean management practices is one of the main focuses of the book. This issue is discussed in different papers by Freyssenet, Camuffo and Comacchio, and it is also explored by MacDuffie. MacDuffie studies the Eisenach plant, GM Europe's most productive plant, and argue that there is a convergence towards a new dominant worldwide management model but at the same time an increasing divergence within country and company. His research is focused on Eisenach's distinctive strategy in adopting lean management system concepts - called "replication". He considered the "template" from which Opel managers brought their experiential knowledge (CAMI and NUMMI), the means of transferring knowledge about the template to the plant (advisors, managers), then the actors (top management, plant management, engineers, workers) who took part in the learning process, and the cultural frame underlying their behavior and solutions adopted.

Many people and organization made this book possible. We express our thanks to conference sponsors: Plastal ZCP, Fiat, Fondazione Carive for their financial support that made the Conference possible, and Fiat Auto for the visit to the Melfi plant in South Italy and the Italian National Council for Research (CNR) for its conference and publication funding.

Our thanks to the conference participants and to the organization staff of the Department that helped making the Conference a most successful and enjoyable occasion.

Furthermore we express our thanks to the Springer staff, namely Mr. Lehnert and Miss. Ellewig for their support.

Finally we would like to acknowledge the fruitful discussion with our many colleagues involved in the study of the Automobile Industry and particularly the members of the International Motor Vehicle Program at MIT and the Permanent Group for the Study of the Automobile Industry and Its Employees (GERPISA) whose researches, suggestions and comments provide a stimulating environment for our studies. We express our thanks as well to colleagues of University of Venice and namely: Enzo Rullani, Sergio Faccipieri, Massimo Warglien and Stefano Micelli.

1.4
References

Altshuler A, Roos D, Jones D (1984) The Future of the Automobile, MIT Press Cambridge.

Freyssenet M, Mair A, Shimizu K, Volpato G (eds) (1998) One Best Way? Trajectories and Industrial Models of the World's Automotive Producers, Oxford University Press, Oxford.

Kochan T A, Lansbury R D, MacDuffie J P (eds) (1996) After Lean Production - Evolving Employment Practices in the World Auto Industry, Cornell University Press, Ithaca.

Shimokawa K, Jurgens U and Fujimoto T (eds) (1997) Transforming Automobile Assembly - Experiences in Automation and Work Organization , Springer, Berlin,

Womack J P, Jones, D T and Roos D (1990), The Machine That Changed the World, Rawson Associates, New York.

2   Industry clockspeed and competency chain design: an introductory essay

Charles H. Fine

2.1
Introduction and motivation

Coping with the dizzying rate of change in the world today consumes much attention of industry and corporate leaders.   Markets, technologies, and competitors all move more quickly than a decade ago and at light speed relative to a century ago. The half-lives of the leading business organizations seem to be shrinking as well, with each technological or organizational innovation unleashing another flood-tide of creative destruction. General Motors, IBM, and Sears each had their day in the sun, Microsoft is having theirs, but history provides one absolute in business as well as politics: All competitive advantage is temporary.

Although many observers have noted the need for organizational robustness in the face of economic turbulence, we have few organizing concepts to guide dynamic business strategy. This paper will introduce industry clockspeed as one such concept. Although the world may be moving faster, not all industries move at the same pace; different industries move at different clockspeeds. Furthermore, inside the firm, the different assets or capabilities may change, grow, or obsolesce at different rates. Therefore a clockspeed analysis of the firm's internal organization may yield useful insights as well.

Using the clockspeed concept as a tool, we will argue that the design and assembly of capabilities in the supply chain is the meta- or inner-core competency on which firms most need to focus. Although the business strategy literature has historically concentrated on the individual corporation as the appropriate unit of analysis, attention has now (appropriately) expanded to the extended organization, i.e., the supply chain--a term we use to mean the corporation plus its supply network, its distribution network, and its alliance network. Much of the supply chain literature addresses supply chain management--the stewardship and utilization of the relevant network of organizations and assets to provide value to some final consumer. However, most of that literature takes the supply chain as given. Just as the manufacturing management community discovered in the past decade the enormous power of the product design activity for leverage in improving product manufacturing performance, the viewpoint here is that thorough consideration of supply chain design can reap enormous advantages for the activities undertaken in supply chain management.

Finally, we will argue that supply chain design ought to be thought of as assembling chains of capabilities for a series of temporary competitive advantages and that these design activities constitute the core of what defines a firm in a dynamic economy. Further, what distinguishes the top-performing firms from the ordinary is the ability to anticipate better where lucrative opportunities are likely to arise and to invest in the capabilities and relationships relevant to exploiting those opportunities. Risk and uncertainty are inherent in investing in hoped-for windows of opportunity, but superior market and technological forecasting ability and superior competency portfolio management are critical since, especially over the long run, fortune favours the prepared firm (Cohen and Levinthal 1994).

2.2
Clockspeed: from fruitflies and infotainment to dinosaurs and airplanes

If an industry has a clockspeed, how might one measure it? Let me suggest several sub-metrics: process technology clockspeed measured by capital equipment obsolescence rates; product technology clockspeed measured by rates of new product introduction or intervals between new product generations; organizational clockspeed measured by rates of change in organizational structures; and "other asset" clockspeeds.

In process technology clockspeed, consider the semiconductor industry as compared to automobiles. A firm such as Intel sinks approximately a billion dollars into a wafer fabrication plant and expects that plant to be essentially obsolete in four years. If they don't get their money out in that time, they will not have the capital to build the next generation of plants. In comparison, a billion dollar engine or auto assembly plant for Ford will be expected to earn significant cash flow twenty years from now. Furthermore, Ford operates very productive twenty-year-old plants with twenty-year-old equipment. Intel has no such relics in its portfolio. Neither Intel nor Ford is necessarily sub-optimizing in this comparison, they merely operate in industries with different process technology clockspeeds.

In the domain of product technology clockspeed, consider the commercial aircraft industry compared with MICE (Multi-Media Information,

Communications and Electronics--sometimes referred to as infotainment). Boeing's rate of (major) new product launches is slightly under two per decade (777 and new 737 in the 1990's, 757 and 767 in the 1980s, 747 in the 1970's). Compare this with Disney studios. In big-release children's animated movies, Disney seems to aim for one new product per year (Beauty and the Beast, Lion King, Pocohantas, etc.). On a corporate basis, a major movie studio may turn out dozens of new products per year, many of which will have their artistic and economic fate sealed in the first weekend after public release. Although these products do have a long tail to their shelf life (Snow White is far older that the 747), Disney's product development teams presumably work on a cycle time geared to the time between new product introductions, a metric that suggests that MICE has a faster clockspeed than commercial aircraft. (More striking, perhaps, is a look at Disney's whole MICE supply chain (no pun intended): the distribution channels and technologies exhibit a very high clockspeed, where business alliances seem to form and dissipate weekly in the contest to see who can win the race for a technology-content package for two-way video, movies on demand, and infotainment.)

Regarding measures of organizational clockspeed, a few suggestive papers are Leonard-Barton (1992) who describes organizational obsolescence as "core rigidities" and Henderson and Clark (1990) who describe how firms might be unable to respond to architectural innovations in their industries due to an organizational bureaucratization around the needs of a previous technological architecture embedded in their principal products. Refining organizational clockspeed metrics will require a more thorough examination of the organizational literature.

Finally, one should consider clockspeed measurement for assets that are not explicitly process technology, product technology, or internal organizational capabilities. Two examples are distribution channels and brand names. Distribution channels such as the Sears catalog, the Walmart department store, and the internet storefront may vary significantly in the rates at which the assets can be constructed and at which they may decay. Similarly, the value of brand names such as Coca Cola soft drinks or Tide detergent may have developed over decades and may be quite durable, whereas Saturn, Lexus, and Yugo automobiles each established a strong brand image in a fairly short period of time.

Two further complexities of measuring clockspeed must also be addressed. First, aside from measuring an industry's mean clockspeed, one must consider its variance. Sturgeon (1996) has observed that both the semiconductor and the circuit board industries are reasonably fast clockspeed, but that microprocessor development has followed a low variance path as predicted by Moore's law, whereas circuit boards were slow-moving until the advent of surface mount technology, which represented a burst of improvement in the technology. Second, industry clockspeed may not be stationary in all (or any) industries. In particular, life cycle effects may exist. One could imagine an industry pattern whereby early bursts of technological discovery generate a fast pace which slows down as the industry matures. Alternately, a slow-moving industry could be hit with an innovation or an increased level of competition which drives the clockspeed up.

Clockspeed may not be stationary within an industry or even well-defined at the industry level since many industries are composites of others (e.g., airplanes are composites of the airframe, engine, and avionics industries, each of which has a different clockspeed). However, the observation that some industries move faster than others has three potential uses. First, clockspeed provides an alternative perspective from which to classify industries. Although a number of industry classification constructs exist, e.g., by capital intensiveness or by concentration ratio, the clockspeed concept suggests a classification that explicitly recognizes the dynamic nature of industry and technology, providing the potential to refine industry-level and inter-industry understanding of Schumpeterian dynamics. Second, the clockspeed concept suggests a cross-industry benchmarking analysis that may prove helpful for firms in designing their extended organizations. Biologists study the "fast-clockspeed" fruitfly species to observe many generations in a short time period and build models of genetic dynamics that are then applied to moderate-clockspeed mammals or glacial-speed reptiles (or perhaps even geologic-speed mountain ranges, although that's probably a stretch). Similarly, industry analysts can observe fast-clockspeed industries (e.g., electronics) to build models of industry dynamics that can then be applied to slower-moving industries such as autos or aircraft. Third, the clockspeed concept may be useful for articulating more concisely the effects of technological or organizational non-stationarities on, for example, a transactions costs analysis of vertical integration (Williamson, 1985).

2.3
A firm is the ability to continually design competency chains for temporary competitive advantage

Fine and Whitney (1996) have argued that the make/buy decision process and the related processes of product development and systems engineering might be thought of as core competencies. That is, the ability to choose intelligently which capabilities should be developed and/or maintained internally may be at least as important as any individual technical capability, for example. This paper extends that argument to suggest that the capability of ongoing concurrent design of products, processes, and the intra- and inter-organizational network of competencies required to deliver value to the marketplace is perhaps THE meta-core (or inner-core) competency above all others. Especially in a fast-clockspeed environment, this ability to develop continually a series of temporary competitive advantages, may be the essence of the firm in a dynamic world. A more dynamic theory of the firm would therefore view a firm as the capability to design and assemble assets, organizations, skill sets, and competencies for a series of temporary competitive advantages, rather than a set of activities held together by low transactions costs, for example.   Such a characterization of firms is consistent, I believe, with the competitive environment of what Goldman, Nagel, and Preiss (1995) have termed "agile competitors and virtual organizations."

2.4
References

Cohen, W.M., and D.A. Levinthal "Fortune Favors the Prepared Firm," *Management Science*, 40 (2), pp. 227-251, 1994

Fine, C. and D. Whitney (1996), "Is the Make-Buy Decision Process a Core Competence?," IMVP Working Paper, Massachusetts Institute of Technology, can be downloaded from http://www.clockspeed.com.

Goldman, Steve, Roger Nagel, and Kenneth Preiss, *Agile Competitors and Virtual Organizations*, New York: Van Nostrand Reinhold, 1995.

Henderson, R., and K. Clark, "Architectural Innovation: The reconfiguration of existing product Technologies and the failure of established firms," *ASQ*, 35, pp. 9-30, 1990.

Leonard-Barton, D. (1992), "Core Capabilities and Core Rigidities: A Paradox in Managing New Product Development," *SMJ*, 13:111-125.

Sturgeon, Tim, *Agile Production Networks in Electronics Manufacturing*.   PhD Dissertation, University of California at Berkeley, 1996.

Williamson, Oliver, *The Economic Institutions of Capitalism*.   New York: The Free Press, 1985.

3   New perspectives on automation

Giuseppe Volpato[1]

3.1
Automation: reference strategy for the automobile industry

Considering the automobile industry evolution of production and assembly technologies, it is easy to ascertain that the pursuit of efficiency/quality binomial growth objective and the reduction of manufacturing time has taken on the form of a systematic replacement process of human labor with machines2. This process has had different denominations: standardization, specialization, production and assembly line organization, mechanization, automation, flexible automation, computer aided manufacturing, and so on. The different names result from the need to highlight the innovative aspect (within the general automation process) and by the succession of technological innovation brought forth, thus the technologies made available. However the basic element, common to all these different phases, can easily be traced to the progressive exclusion of human labor in favor of machine power and activities (Hounshell 1987; Abernathy 1983; Hsieh et al. 1997). This was first seen in simple manual operations and it then extended to operations where the intellectual nature of the job became more important, but the objective was to replace man with machines.

The concurrent reasons determining this evolution are numerous and have different degrees of importance according to the period of time considered and characteristics of the working conditions (economy, technology, and social) of the automobile manufacturers in the various countries (Freyssenet, Mair, Shimizu

[1] The author would gratefully aknowledge CNR (Italian National Research Council), Murst (Ministry of University and Scientific and Technological Research) and IMVP for their research funding
[2] On the strategic role of automation and automation industry as well see: Rosemberg (1963).

and Volpato 1998). For example, in certain moments the replacement process, more generally referred to as "automation", was mainly driven by the need to reduce production costs, while in other moments by the need to reduce human intervention in dangerous and harmful jobs or to obviate the decrease in availability of desired labor force.[3]

Recently, there are phenomena significantly modifying this orientation in the automobile industry. The trend towards automation, machines replacing human workforce, has not stopped. But it is entering a new phase, implying a deep reorientation of the automobile industry management. This means the entire present and prospective reference schemes, in which the defining strategies of production and assembly processes materialize, are influenced by an intense redefining phase. Scholars and business managers are reexamining not just the way automation could be implemented, but also the way automation is conceived and its potentiality and limits evaluated. So it is necessary to analyze the main aspects characterizing the mutation process going on in the international automobile industry, in order to provide an adequate reference scenario in which the most recent trends can be interpreted by the theoretical debate on automation and choices can be made by the automobile manufacturers.

While keeping clarity and conciseness, the analysis could group the changing factors of the reference scheme into three main categories. They have a strong degree of mutual interaction, but they can be considered separately:

1. repositioning of the international automobile demand;
2. offer globalization, in its diverse meanings;
3. a new division of labor between automobile manufacturers and component suppliers.

The aim of this paragraph is to analyze these phenomena and obtain indications to modify perspectives and automation choices during implementation phase by automobile manufacturers.

---

[3] On this subject see the different experiences of car manufacturers presented in Shimokawa et al. (1997), and in particular: Mishima (1997), Niimi and Matsudaira (1997), Wilhelm (1997), Decoster and Freyssenet (1197), Camuffo and Volpato (1997).

3.2
Mutating factors of the automation reference scenario

3.2.1
Repositioning of the automobile demand

As known, the growth of motorization was geographically different in time, according to the buying power level of populations, though it privileged the highly industrialized countries: United States of America, Western Europe and Japan. Today, these three large areas have reached levels of motorization close to saturation, since the number of vehicles circulating per one thousand habitants is estimated around 400 units in 1997, and the number of vehicles per one thousand habitants around 450 units, with a maximum peak in the United States of 520 vehicles per one thousand habitants. In other words, almost every household resident in these three areas has at least one car and the automobile demand is nearly exclusively a demand for replacement. This does not mean that the annual vehicle sales in these areas is to decrease but, besides the predictable cyclical fluctuations typical of mature markets, the growth rate of this market group can only be very moderate and likely linked to the important technological innovations of the conventional vehicle (Otto and Diesel cycle endothermic engine) as well as state incentive laws for city use of zero emission vehicles and "hybrid" vehicles.

In other words, the future growth of the automobile demand will be determined mainly by the dynamic motorization process in eastern European, South America and Asia (in particular, India and China). These are highly populated areas4 in which development requires and promotes diffused motorization. It is sufficient to consider the present motorization levels of the developing economies to understand the existing margins of growth. This emerges from the latest statistics available, even if not recent, and indicates the way for the next evolution. As shown, on Table n.1, in 1989 only Western Europe, North America, Japan, and Australia could count on a vehicle circulation level greater than 300 units per one thousand habitants. In 1995 the absolute values of motorization did not seem to change much, but in these six years the percentage rate of variation clearly shows global repositioning of the motorization process. Even hypothesizing that the future motorization rate in these new areas could not exceed half the value presently reached by industrialized countries, there is a significant demand for cars and trucks.

---

[4] In 1995 the population in Eastern Europe was over 400 million, in Mercosur (Argentina, Brazil, Uruguay and Paraguay) was almost 200 million, while China and India together exceeded two billion habitants.

Table 3.1- Motorization levels in the world (cars and trucks per 1,000 habitants)

Geographic Areas	1989	1995	Diff%	1989	1995	Diff%
	Cars	**Cars**	**Cars**	**Vehicles**	**Vehicles**	**Vehicles**
Western Europe	345	379	9,86%	387	432	11,63%
Eastern Europe	65	92	41,54%	94	122	29,79%
North America	399	362	-9,27%	523	530	1,34%
South America	63	70	11,11%	84	89	5,95%
Africa	13	15	15,38%	20	23	15,00%
Asia	16	24	50,00%	32	37	15,63%
Asia (without Japan)	6	10	66,67%	11	16	45,45%
Oceanic	368	373	1,36%	472	496	5,08%
World	83	88	6,02%	110	119	8,18%

Source: Our formulations on national statistics of various countries.

Table 3.2 – Automobile production in the World (estimates) - Millions of units

Geographic Areas	Cars				Tot. Vehicles			
	1995	**2000**	**2005**	**2010**	**1995**	**2000**	**2005**	**2010**
Western Europe	13,5	13,9	14,6	14,5	15,3	16,0	16,9	16,9
Eastern Europe	1,5	2,9	4,0	4,9	1,8	3,4	4,6	5,7
North America	8,4	8,9	10,3	10,7	15,3	16,5	18,5	19,3
South America	1,6	1,9	2,3	2,7	2,0	2,4	2,9	3,3
Africa	0,3	0,3	0,4	0,5	0,5	0,6	0,7	0,7
Asia	10,9	13,2	14,7	16,2	15,7	20,1	23,1	26,1
Asia (without Japan)	3,3	6,3	8,4	10,4	5,5	10,8	14,0	17,3
Oceanic	0,3	0,4	0,4	0,4	0,4	0,5	0,5	0,5
World	36,5	41,5	46,7	49,9	51,0	59,5	67,2	72,5

(*) The world total is slightly overestimated due to CKD assembly count in SouthAmerica, Africa e Oceanic.

Source: Our formulations on LMC, DRI estimates

According to the estimates made by the different research centers, the entire world offer should reach 60 million vehicles by the year 2000, of which 70% cars and exceed 70 million by 2010. These estimates do not always conform to geographic distribution of production increments, but the global values substantially converge.

The main thing to consider is that the various research centers agree on predicting a leveling out of production in two of the traditional automobile manufacturing areas: Western Europe and Japan; while North America should continue to increase its offer, due to the rise of light trucks, including the constant growth of pick-ups, as well as jeep and mini-van models, which are part of this class of vehicles. As far as the developing countries are concerned, strong production increments are expected first of all in Asia (Japan not included) and then in Eastern Europe and South America.

The evolution of the automobile manufacturing scenario will be characterized by a strong change of the market areas importance in the next ten years. It is unthinkable that this motorization process can take place through vehicle exports from the present production areas with strong automobile manufacturing specialization. Aside from political needs, there are sufficient economic reasons to strengthen development of production capacity in countries having an additional vehicle demand. It is a current process, as shown by the statistics on car and truck manufacturing. In 1985 the total number of vehicles manufactured in Western Europe and North America was over 64% of the total world production, in 1995 it fell to 60% in these two areas. South America and Asia (without Japan) had the strongest growth rate since new productive capacity plants practically doubled the vehicle offer in these areas between 1985-1995 (Humphrey et al. 1998).

The new phase will symbolically open in the year 2000, but it has already begun. It is characterized by a greater geographical repositioning process of the motorization demand as well as the consequent productive localization towards countries exceeding the richness and social organization threshold, so to start a mass motorization process. The speed of this change is not easy to predict, as it depends on the solidity of economic and industrial expansion these countries will show. In the past, these countries have suffered strong reactivity to economic fluctuations. According to esteemed hypothesis, the global amount of registered vehicles by 2005 should be over 67 million.[5] If the three large traditional areas (called "strong areas") of motorization will absorb approximately 65% of the automobile manufacturing production by 2005, the demand for vehicles in new areas of the mass motorization process will be almost equivalent to 23 million vehicles; almost double the 13 million vehicles registered in 1995. In other words the number of vehicles absorbed by countries presently at the margins of the motorization process will soon be a little less that half the present total absorption of the world automobile market. Therefore, this is a real revolution of the global automobile industry. It is an even more visible transformation, because it will occur in a short time, compared to the first phases of motorization in the three strong areas.

The most advanced example of this process is represented by the Mercosur countries (Brazil, Argentina, Uruguay and Paraguay). In particular Brazil and

---

[5] 71% should be made up of cars, which is equal to a little over 46 million units.

Argentina polarize investments regarding the establishment of new productive capacities from the international automobile industry.6 Due to these investments, some of which are still being implemented and will have effects in the next few years, production of vehicles in Mercosur went from one million units to 2.5 million units between 1990 and 1997 (Figure 3.1).

This repositioning process of vehicle demand brings a wide reorganization of the international automobile industry, commonly known as globalization: a term frequently used when describing the future strategies of automobile manufacturers. It is a "fashion" concept giving a modern and competitive aggressive aspect to the automobile company policies. The "globalization" concept has been given many meanings which should be kept apart to avoid misunderstandings and ambiguity. Therefore, distinct forms of globalization are considered each time in reference to the geographic area, production and product.

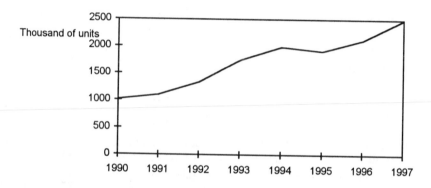

Source: Our elaboration on Anfavea data

Figure 3.1 Vehicle Production in Mercosur

3.2.2
Geographic globalization

The term "globalization" means a planetary expansion of company production presence, so the closest meaning has a geographic reference. A company operating within the automobile *filière* as vehicle assembly or components manufacturer, has a "global" nature because it is present with production (as well as trade) in all main markets7. This is the most simple and intuitive definition of the first globalization concept. It is different from the previous phase generally

---

[6] On this subject see the contributions presented in Arbix and Zilbovicius (1997).
[7] See: Actes du Gerpisa n.18, (1996) and n.22, (1998).

called "internationalization" which refers only to the commercial presence through vehicle export from the domestic market. In the 1980's, automobile companies and manufacturers were defined "global" in the strict sense as previously mentioned. General Motors, Ford, Toyota not only had a commercial presence in all main automobile markets, but they also had production and assembly plants in several countries.

However this first form of globalization is characterized by a low levels of mutual integration among the various productive units in the world. Its diffusion originates from different needs, often contingent, and it is not the result of an organic development plan at a worldwide level. For example, the international expansion of the automobile companies was very different whether it was oriented towards highly industrialized countries, such as American companies to Europe and the European companies to North America, or towards developing countries. In the first case the main need was to plan and manufacture specific vehicles for the selective and sophisticated market demand of these areas (Abo 1994). In the second case, from the automobile company's point of view, the implementation of productive capacities abroad represented a second best when compared to an international expansion given exclusively by commercial trade flow.

For example, from the second half of the 1950s, a slight and generalized growth of vehicle demand could have brought heavy trade imbalances for the non manufacturing countries. Therefor, a significant number of governments opted for tariff barriers to dramatically reduce imports. Among the countries that decided to go in this direction are Spain, Brazil, Argentina, Mexico, Australia. Automobile manufacturers had to chose between consistent involvement in direct investments or exit those markets. The prevailing choice was to stay. This determined a proliferation of oversized plants when compared to the demand of the single markets, but clearly undersized when compared to the potentiality needed to have production costs at the same level of efficiency standards of the primary markets (Volpato 1983). The most logical choice would have been to program a network system of large specialized plants for each continental area from which the various markets are supplied by forms of logistic integration. This solution was not feasible faced to tariff protection policies of each country and the extreme complexity due to absence of communication means only recently made available. The result is an international expansion, scattered here and there, derived from undersized units along with low integration. Today, a strict geographic definition of globalization is not considered sufficient and leaves space for a higher meaning of globalization, referable to as "production globalization", to distinguish it from the simple geographic nature of the first phase.

3.3
Production globalization

A more important and complex globalization stage is surmounted by the pure geographic dimension and reaches a form definable as "productive globalization". This implies not only an important level of specialization of the various production units but even a sophisticated logistic integration system of production plants and final assembly centers. It is necessary to create adequately sized manufacturing plants for highly productive systems and machinery and for complete economies of scale use. This requires the implementation of plants that are oversized to serve the single markets. It is therefore necessary to create a triangulation process to send the manufactured parts to the final assembly plants and finished products to the market. In other words, production globalization implies: a integration network project, specialized production centers, and logistic capacity to fluidly and economically manage the great flow of materials and finished products, as required by this scheme (Womack, Jones and Roos 1990; Mair 1994).

This is a delicate and complex phase, as it is still hindered by a series of limits linked to the technical constraints as well as juridical, tariff, and market aspects. The wellknown limits imposed by countries on the domestic content rate of automobile production (nationalisation) are often in contrast with the system specialization criteria. In addition, there are contradictions among the different types of strategies the manufacturers would want to use. For example, certain aspects of the production globalization scheme oppose the just-in-time supply, because the triangulation system of components flow over large geographical areas does not allow forms of pulled services from the final assembly line as well as stock elimination, which is a key objective of automobile companies. Of course, the term "component part" refers to a very complex universe of products with different technological, productive and economic specificity. So the economic advantages of the solution based on specialized production centers should be applied to certain components (engine, transmission, electronic systems, etc.) characterized by high added value, greater standardization rate, greater economies of scale and heavy investments on equipment. On the contrary, the solutions based on vicinity of productive and assembly centers should certainly be more important for the other components (seats, dashboards, bumpers, etc.) having a high range of variability and which transport costs strongly influence the intrinsic value.

This second phase of globalization has already begun and can be effectively completed by a further globalization level of product planning.

3.4
Product globalization

In abstract, the simplest way to rapidly implement a production globalization plan, is to make a car that can be manufactured and sold anywhere. This concept is traditionally summarized by the term world car, meaning a car suitable for markets worldwide (Maxcy 1981; Sinclair 1983). The automobile industry has already seen models similar to what is considered a world car. The first case was the Ford "T", presented in 1909, which had a very high potential diffusion, but only in part achieved because a series of strong tariff limits were introduced after World War II. Another similar case of the world car concept is the Volkswagen "Beetle", designed by Engineer Ferdinand Porsche in the 1930s, which manufactured over 20 million models starting from 1945. In a certain sense, these vehicles are the "purest" cases of world cars as they were manufactured in a very standard way. The products were made in only one model and there was no possibility to choose among different versions, and the evolution, during their product life-cycles was modest compared to the starting scheme.

After World War II, automobile companies got far from standard schemes, mainly in the United States, to look for a broader range of versions and models with shorter renewing cycles. The fuel crisis in 1973 and 1979 launched the world car concept again, in order to reach vast economies of scale, reduce production and distribution costs, as well as re-launch a demand heavily influenced by recession. Even in this case, Ford leaded the way with the "Fiesta" model and above all the "Escort" model. The latter was manufactured both in Europe and the USA, even if a modest level of component parts were in common.

Also General Motors in the 1980s looked for a stronger standardized form, but differently oriented, compared to Ford. Ford tried strong forms of standardization which can be summarized as "manufacturing the same vehicle in many countries and selling them in many markets". Due to its history, General Motors was more incline to manufacture cars that could be differentiated for the specific images of its divisions (Chevrolet, Buick, Cadillac, Oldsmobile, etc.) and for the requirements of different markets. This policy was implemented by designing a limited number of platforms, motors, and transmission-suspension groups to be used by a wide range of products. This can be considered a useful and interesting experiment. But it has not yet reached the results hoped for because the various market consumers express differences that penalize marked forms of standardization and the technical and operational problems of this design turned out more difficult than expected.

For parts not influencing product image the commonality way seems to be a mandatory step for all car manufacturers and most of all, for those companies with a plurality of brands, and product/market combinations such as General Motors, the VAG group, FIAT group, etc. All main automobile groups have programs to reduce the number of platforms and direct suppliers. As far as supplier reduction is concerned it is a widely generalized phenomenon but not yet complete. For instance, the movement towards this direction is followed not only by automobile groups like Fiat, Renault and PSA that have already showed a strong orientation towards these forms of commonality, but also by automobile companies once contrary are now showing a similar propensity.

It is clear that productive globalization is more rapid and efficient as the analyzed product globalization form progresses. Anyway it should not be forgotten that the various market areas have so many different conditions and specificity that a simple standardization strategy shall face inevitable and insurmountable market difficulties. A more complex but feasible form is given by the use of standardization for certain vehicle functional systems, which are less detectable by the final consumer and have significant economies of scale.

Nonetheless, a new product globalization concept is emerging. It embodies a more abstract and sophisticated form of commonality than that based on standardization of single component parts. This new and different form of the globalization concept is not evident and may more easily be defined by using an example. A kind of strict commonality is represented by the use of the same suspension systems for different platforms. However this form of commonality has the disadvantage of manufacturing vehicles suitable for countries with good road conditions (USA and western Europe), but which will result unsuitable for those with poor road conditions, or viceversa. To overcome this problem means to design platforms and bodywork that allow the of use alternative suspensions, having different design and weight as well as a series of common design and production specifications (same attachments, same overall dimensions and assembly methods), to favor interchangeability of the various sub-systems composing the car unit. In other words, by using the term "sub-system" to indicate a functional element of the whole vehicle, such as the engine, the transmission, the suspensions, the braking system, the air conditioning, the dashboard instruments, the exhaust system and with the word "module" an alternative version of the same sub-system, this means it is possible to manufacture a variety of different vehicles by simply combining a plurality of available modules and ensure the same sub-system functionality. Here the meccano-play logic finds full application, in other words it would be possible to articulate the vehicle offer by using different modules for the same sub-system, based on the different needs of each market.8 If two vehicles are to be sold on very similar markets, the maximum level of commonality can apply: same vehicle (except for usual color, interior, and optional differences) made by using the same modules for each sub-system. Should the vehicles be sold on non similar markets, the common forms are inferior. They are maintained for platforms and assembly, as the single functional sub-systems are made up of modules with different but interchangeable performances.

---

8 In the hypothesis of full use of the meccano-play concept it is possible to implement a number of different cars equal to the multiplication of the modules available for each sub-system.

3.3
The new labor division between automobile companies and component suppliers.

3.3.1
Growing specialization of the automobile companies.

Intensification of competitiveness among automobile companies has lead to deep changes in the division of labor in the automobile *filière*, in the last ten years. The Fordist strategy, based on clear vertical integration of automobile companies has been over turned into a policy delegating suppliers a growing range of jobs, such as production as well as development and design of innovative components. Automobile companies have progressively gone from designing the entire car and outsourcing only the production of components, to the present situation based on the external delegation of design, development and production of the new components. A co-design relationship is established for parts having the typical distinctive aspect of the car brand image and performance.

The automobile companies objective is to reduce the area of activity and increase potential flexibility, so to vary productivity volume and respond more rapidly to technology changes of the automobile industry. For the automobile companies, the aim is focused on leaning occupational levels and investment in order to help the recovery of financial resources to concentrate in globalization initiatives, and to have higher return on investments (Camuffo and Volpato 1997). Nevertheless it should be mentioned that the "shortening" maneuver of the span of activities presents risks for decreasing the contractual power of the automobile companies towards suppliers and prospectively reducing the ability to adequately personalize its products. However, looking at the huge amount of capital necessary in the next few years to enter emerging markets, it is difficult to imagine alternative ways of proceeding. Time and intensity of this transformation can be discussed, but the method has already been decided.

3.3.2
Balancing of supply along the *filière*: concentration and hierarchical system

As described above the automobile companies repositioning suggests a mutual change by the rest of the *filière*. This transferring process is heading towards completion. It had an important impact and was a significant transformation in the automobile *filière* from a flat system to a hierarchical organization.

Obviously this reorientation could not involve all the previous suppliers. Until component development and planning was the contractor's job, production could be distributed among different mutually competitive suppliers. As outsourcing requires consistent investments in research and specialized equipment , it could not be delegated to each single supplier. Probably not all had adequate technical, financial and managerial resources to rapidly respond to the needs of automobile companies. In addition, conveying production to fewer suppliers was necessary so the supplier could make all necessary economies of scale and amortize the heavy investments needed by this significant transformation. The transformation is going from a manufacturing process based on the car maker to a manufacturing process divided according to vehicle functional systems (engine, transmission, suspensions, brakes, headlights, electrical system, etc.) and given to functional specialists appointed to produce and integrate a vast number of single components in an accomplished sub-system. Single component suppliers can no longer directly refer to the automobile manufacturer, but they must intensify relations with companies nominated to ensure functionality of the subsystems part of the vehicle. It goes without saying that this process of hierarchical organization considers the complex differences characterizing the single vehicle components. For example, if a second level supplier manufactures non critical and relatively standardized components, connections with the automobile companies completely disappears. If there is a complex, costly and personalized component, compared to the functional characteristics of the vehicle to manufacture, exchange of information with the manufacturing company is required by the co-design relationship. The result is that all the car manufacturing companies have greatly reduced their number of direct suppliers.

Certainly this transformation is not over. On one side, there are still margins for further outsourcing, which the automobile companies do not intend to bring forward at the moment; because an adjustment phase to suitably assimilate this first period of deep and frenetic transformation is necessary; and automobile companies wish to maintain a number of suppliers for the same type of functional system and rotate the contract supply, according to their vehicle range and model replacement needs. This is a way oriented to:

- protect themselves from the growth of sub-systems suppliers contractual power;

- access all opportunities offered by technological innovation;

- consider the degree of product nationalization (local content) in the different countries[9], the deployment of advantages linked to complex set of variables (past experiences, common language and culture, geographic dislocation of various

---

[9] Many automobile companies had and still have important public financing linked to reaching and keeping certain local content rates. This degree of nationalization is the ratio between the added value obtained in a country and the vehicle commercial value.

plants[10] and so on), and the risks due to change of currency and forms of monetary compensation among the areas where components are obtained from suppliers and manufactured vehicles are sold.

On the other hand, the automobile companies plan to pilot this process and favor emerging dispositions more compatible with their long term objectives. For example, if a automobile company plans to make important investments for assembly in a country which previously was not served much, it progressively develops relations with subsystem suppliers of the objective market. These are able to follow and satisfy its needs. Another example the habit of many automobile companies to indicate the person responsible for functional system components that the second level supplier must refer to. In this case, not only an economic calculation is at stake, but also a complex system of evaluating general policies, such as the need to mediate preferences expressed by the technical functions of the automobile companies. This is typically represented by the design and production areas traditionally privileging technology and logistics reliability of supplies and purchasing, when evaluating   the possibility of reducing outsourcing costs at the beginning.

Consequently, even if the hierarchy organization phase of the *filière* is not yet complete, its effectiveness is unquestionable and, most importantly, it has started a further reorganization processes. In other words, the present phase may not be considered a balancing phase as it is still being completed. As competition is not destined to decrease and, as far as the advantages, the hierarchy organization process will reach its completeness and diffusion, reducing its creative potential of economies, the first steps of a new reorganization phase in the automobile *filière* can be seen. This considers a greater concentration phase between first level suppliers (subsystem suppliers) and their wider based organization so to transform them into almost equal interlocutors of the automobile companies.

The objective of this complex reorganization process is to research competitive advantages given by cost savings per product unit, made feasible through the new order, and the innovation rate (Nishiguci 1994). In the next few years, there will be a wider rationalization imposing a different order in the automobile *filière*. The reasons for a further rationalization comes from the efficiency gained during this time, that does not allow the automobile companies to balance economic situations because they are still under strong competitive pressure. Due to excess productive capacity in North America and Europe, the rationalizing pressure within the automobile companies and towards the upstream *filière* decreased and will continue to in the future.

Financial analysts highlight that the average profitability rate of the primary component suppliers was higher than the average of the automobile companies in the last few years. Therefore, new requests are emerging from the contractors to

---

[10] Since the presence of component suppliers may considerably vary from country to country, the ways to obtain *just-in-time* supply can cause to make different choices according to the location of the various assembly plants.

increase service guaranteed by the first level suppliers to the manufacturers. This is both in terms of project coordination of first level suppliers in relation to second level suppliers and in terms of productive coordination.

The process begins when the automobile companies identify functional sub-systems in which the vehicle can be sectioned. Therefore the company recognizes the sub-systems judged to have a substantial role for the company's image and competitiveness to directly protect through large investments in research and development. It may seem like a simple operation, but it requires to make difficult and risky choices. First of all this is the eve of another emission of electronic components in vehicle functioning which could make design and productivity know how, from the past experiences of the automobile companies, rapidly obsolete. The choice depends on identifying distinctive elements that are in the future critical aspects of brand image. This means, for example, functional systems with greater impact on model safety could receive from a brand like Volvo more attention than those connected with a higher engine performance impact which could receive maximum attention from a brand like Alfa Romeo.

Instead, for less critical sub-systems, the direction taken transfers coordinating responsibility to the supply system.

3.3.3
Global sourcing

The last and probably most difficult stage of the component supply system reorganization is in the implementation of global sourcing. Car manufacturers implement a global sourcing system when they are able to buy parts by choosing from offers worldwide. This also requires a certain organizational structure allowing to monitor a wide range of potential suppliers in all the continents, as well as evaluate systems and systematically control the supplier's performance (effective and potential).

Nevertheless, global sourcing systems are sometimes contradictory when compared to the broadening of services (especially logistics) from the supply system. While the global sourcing system hypothesizes varying the source of supply, according to each vehicle renewal, maintenance of high logistic services – lack of stock, pulled supply, sequential component delivery according to assembly line order (in a word: just-in-time) – requires suppliers that are close to the assembly line. A large amount of fixed investments are involved and cannot be recovered during the few years of the model life cycle. If just-in-time supply is implemented with certain suppliers only for a few years and then replaced by a competitor heavily reducing component costs, it is clear that economic sustainability decreases. The final organization model will trade off these two modalities according to the specificity of the supply. For products used indifferently, the automobile companies generalize a global sourcing system. For single vehicle model components, a co-design relationship and just in time supply prevail. In any case, this analysis highlights the subsequent transformation of the automobile *filière* and the supply relationship will be seen in the next few years.

3.4
The impact of the new automation scenario

3.4.1
New localization and automation

The previously analyzed and summarized phenomena - demand repositioning, various forms of globalization and new labor division in the automobile *filière* - show a strong impact on the choices regarding the degree of automation introduced in the new plants. This is not an unambiguous and generalized impact. It has a different importance and valence according to the plurality of variables, which should separately analyzed, as there are:

a) factors putting substantial generalized pressure to reduce propensity to invest in automation;

b) factors putting pressure to reduce selective investments in automation for special types of systems or productive activities;

c) factors putting pressure on special automation system set up.

The fact that new manufacturing plants for parts and final vehicle assembly are mainly in the developing countries, is a factor determining a less convenient use of highly automated systems. Lower cost of labor in these countries compared to mature motorization areas is the most known factor. Moreover, the savings on personnel costs induced by automation will take more time to balance out the additional investments made for more sophisticated systems. However this aspect seems to become less important because the wage growth in the economy of developing countries reduces the previously existing differences, and most of all, because the automobile companies are strongly influenced by other types of limits. Coverage for greater investments on higher levels of automation requires to guarantee system saturation and limit breakdowns or system malfunction to a minimum. These are two conditions difficult to obtain in the developing countries. First because a strong conjectural variable of the automobile demand is linked to the strong variability of the economic cycle. Second, maintenance and repairs of highly automotive systems is more costly and difficult in these countries compared to the three highly motorized areas. It is not the higher cost of the systems that discourage adopting sophisticate automation technologies but the difficulty of having highly reliable assistance for immediate intervention. In the future the same process of opening new productive systems will favor positioning of automation supplier companies in these countries, but now the situation suggests a certain caution. In any case, another important element favoring the limitation of labor replacement with forms of automation are linked to institutional types of conditioning.

A first institutional conditioning is represented by "nationalization" laws on vehicle manufacturing. In other words, these countries recognize the title of "national" product only when productive activity is implemented domestically in a certain percentage of the final value of the car. This percentage is likely to be reached (easier to demonstrate) when production and assembly systems have a high incidence of labor costs compared to fixed costs of the system.

A second conditioning comes from the incentives of central and regional governments of each country for implementing new industrial systems. To attract new plants in their area of competency, these regional governments often have aggressive policies when helping companies. Traditionally, the importance of the localization incentive is commensurate to direct occupation generated by the new system. So the convenience of the automobile companies to limit automation rate of new plants is clear.

There is no doubt the new repositioning process of automobile demand and "production globalization", as defined above, puts pressure to limit the quantity of importance of automation introduced in the new production and assembly systems. These limitations of automation introduction rate do not operate in the same way towards the various forms of automation and different types of systems. On the contrary, they have a strong discriminating effect which should briefly be explained. The discriminating selective mechanism of advanced automation forms originates from the fact that automation can be finalized to a plurality of objectives, grouped into three categories:
    - automation aimed at increasing system efficiency.
    - automation aimed at increasing product quality (in a wide sense);
    - automation aimed at eliminating human intervention in difficult and harmful operations.

3.4.2
Quality standard and automation

In reference to the first type of automation, as the choice is based on an exclusively economic calculation, the considerations made on the general tendency to limit the degree of automation, compared to the most industrialized areas, are still valid. Instead the second type of automation needs a different explanation, as the qualitative improvements of production is an impelling objective also for production in these countries. This occurs because the same local demand is a more and more sensitive to high quality products and therefore it shows a greater availability in recognizing the premium price to the quality product than in the past. Moreover, within the globalization strategy based on the world car, automobile products manufactured in the developing countries are often exported to the earlier motorized markets, where to reach high quality levels is a must. Consequently, productive technologies can be based on different levels of automation, as long as it does not imply quality differences perceivable by the final consumer, if plants are located in earlier or recently industrialized countries. This means, in setting up new plants in emerging markets, there is a tendency to have more automated equipment within the stages with greater technology capacity (compared to human workforce) to implement higher quality standards. If the productive operations in the automobile *filière* are grouped into five large areas - machining, stamping, welding, painting, and assembly – there is a tendency to have higher automation levels for welding and painting. In general, the access to automated systems for welding gives higher quantity standards through the use of robots at welding points (absolute constancy of number of points and positioning) as well as for the body geometry (by special positioning masks). When implementing new plants, there is a greater propensity to invest in automation in these two productive areas, rather than in machining, stamping, and assembly in which manual operations can reach excellent quality standards.11 Surely, the general conditions of economic convenience do not lessen, therefore automation is selective also for welding and painting. Welding for assembly of the chassis, that is the general structure of the body, ensure the passive safety characteristics of the vehicle as well as necessary torsion rigidity. Instead for the inner and less accessible body and the mobile parts (doors), welding continues to be manual. This job is often delegated to external companies which work small pieces of sheet metal. The same considerations are made for painting operations, where automation is used, for example to top coat the external part of the body.12 Traditional methods still prevail for internal painting.

Considering automation forms aimed at dangerous and harmful operations, the choices of manufacturing processes are the result of the preferences of the automobile companies as well as the specificity and intensity of the unions. It is not possible to identify a clear trend since the unions are also in difficulty when the choice to reduce the number of risky operations penalizes occupational levels. Automation is systematically present during the coupling phase in final assembly of body with engine-transmission–suspension group and during the electro-coating process. Traditional methods still prevail during the other painting phases, even if the harmful effects are considerable. In addition, automation assist forms, which developed in modern plants localized in the earlier industrialized areas, are lacking.

---

[11] Inserting robots in assembly operations, especially when gluing the windshield wippers, is an exception to this general line.

[12] It is interesting to note that the trend is to outsource painting to companies specialized in preparing paint (for example, PPG) through leasing the painting system within the plant.

3.4.3
New distribution of work and automation

The new distribution of jobs along the automobile *filière* between component suppliers and automobile companies, highly specialized in final assembly, will have repercussions on automation. The transformation has not yet been implemented and can only be hypothesized without .empirical comparisons. Presently, the most possible automation level transformation will come from the continued cutback of the final assembly. This to accommodate a small number of stations in which entire sub-systems are pre-assembled and pre-tested by suppliers and structured in "modules". Final assembly is made up of large size movements of few complex modules which include a great number of elements. From a quantitative point of view, the operation is becoming more simplified, but from a qualitative point of view the process is increasing its complexity, because delicate manipulations are needed and can be guided only by a high resolution vision. In this operation man excels the machine. Presently, there is not a robot system able to compete with the sensorial sensibility of the hand and the visual precision of the eye.

Moreover, the fact these modules are necessarily cumbersome and heavy must not be forgotten. It is unthinkable that the person responsible for module insertion has also to implement movement. The most probable hypothesis is to organize assembly based on the presence of both "machine" and man. The machine's job is to move the single modules and bring them close to the final position. Man must implement all necessary module connections and make all the micrometric adjustments of the parts connected, while verifying the validity of the final result. As explained this is a sort of identikit of the above mentioned automation assist. Reasonably considering the work distribution between manual and automated system, in the future there will be no further increments of the levels of assembly automation by the automobile companies. The diffusion of automation assist should prevail, as it congenial to the future of the assembly line disposition.

Different considerations are necessary for the component manufacturers, as far as the first tier are considered. They will be forced to change from suppliers of separate parts to complex module suppliers. The largest part of assembly will be upstream along the *filière* and impose a large labor reorganization of the suppliers, because the importance of assembly will greatly increase compared to the past. Aside from the automobile companies, here assembly is rather simple and standardized and largely implemented by automated robots. Two important factors play in favor of this evolution. On one hand, the car will have less mechanical components in favor of more electronic components. This element helps to diffuse automated assembly equipment. Today this is seen in electronically based assembly equipment, like computers. On the other, by transferring upstream part of the module assembly to fewer first level suppliers, it should favor utilization of economies of scale by systems with a high automation degree compared to the present assembly line used by automobile companies. This is a result of the greater simplicity of assembly implementation (module assembly, i.e. a vehicle sub-system, instead of the entire vehicle) as well as the difference of certain module elements (ex. use of electronic cards) not impeding assembly of modules for different vehicles of same brand or other brands. Therefore the incentive for automation will be more potential.

In conclusion, the present transformation phase of the international automobile industry can be considered as a model of profound changes in the productive automation field.

3.5
References

Abernathy, W.J., 1983, The Productivity Dilemma: Roadblock to Innovation in the Automotive Industry, John Hopkins University Press, Baltimora.

Abo, T. (ed.), 1994, The Hibrid Factory – The Japanese Production System in the United States, Oxford University Press, Oxford.

Actes du Gerpisa, 1996, Between Globalisation and regionalisation, n.18, Novembre, Paris.

Actes du Gerpisa, 1998, *Internationalization of Firms: Strategies and Tajectories*, n.22, February, n.22, February.

Arbix, G. and Zilbovicius, M. (eds.), 1997, De JK a FHC – A reinvenção dos carros, Edições Sociais Ltda, São Paulo.

Camuffo, A., and Volpato, G., 1997 a, Building Capabilities in Assembly Automation: Fiat's Experiences from Robogate to the Melfi Plant, in Shimokawa et al. (eds.), 1997.

Camuffo, A., Volpato, G., 1997 b, Nuove forme di integrazione operativa il caso della componentistica automobilistica, F.Angeli, Milano.

Decoster, F., and Freyssenet, M., 1997, Automation at Renault: Strategy and Form, in Shimokawa et al. (eds.), 1997.

Freyssenet, M., Mair, A., Shimizu, K., Volpato, G. (eds.), 1998, One Best Way? – Trajectories and Industrial Models of the World's Automobile Producers, Oxford University Press, Oxford.

Hounshell, D.A., 1987, From the American System to Mass Production – 1800/1932, The Development of the Manufacturing Technology in the United States, John Hopkins University Press, Baltimora.

Hsieh, L-H., Schmahls, and T., Seliger, G., 1997, Assembly Automation in Europe – Past Experience and Future Trends, in Shimokawa et al. (eds.), 1997.

Humphrey, J., Mukherjee, A., Zilbovicious, M. and Arbix, G., 1998, Globalisation, Foreign Direct investment and Restructuring of Supplier Networks: The Motor Industry in Brazil and India, in Kagami et al. (eds.), (1998).

Kagami, M., Humphrey, J., and Piore, M. (eds), 1998, Learning, Liberalisation and Economic Adjustment, Institute of Developing Economies, Tokyo.

Kleinwort & Benson Research, 1996, Component Suppliers Review, December, Londra.

MacDuffie, J., P. and Pil, F., K., 1997, From Fixe to Flexible: Automation and Work Organization Trends from the International Assembly Plant Study, in Shimokawa et al. (eds.), 1997.

Mair, A., 1994, Honda's Global Local Corporation, Macmillan, London.

Maxcy, G., 1981, The Multinational Motor Industry, Croom Helm, London.

Mishima, Y., 1997, Present State and Future Vision of the Vehicle Assembly Automation in Mitsubishi Motors Corporation, in Shimokawa et al. (eds.), 1997.

Niimi, A., and Matsudaira, Y., (1997), Development of a New Vehicle Assembly Line at Toyota: Worker-oriented, Autonomous, New Assembly System, in Shimokawa et al. (eds.), 1997.

Nishiguci, T., 1994, Strategic Industrial Outsurcing. the Japanese Advantage, Oxford University press, New York.

Rosenberg, N., 1963, "Technological Change in the Machine Tool Industry", in Journal of Economic History, n.4.

Shimokawa, K., Jürgens, U., Fujimoto, T., (eds.), 1997, Transforming Automobile Assembly – Experience in Automation and Work Organization, Springer-Verlag, Berlin.

Sinclair S., 1983, The World Car – The Future of the Automobile Industry, Euromonitor, London.

Volpato, G., 1983, L'industria automobilistica internazionale, Cedam, Padua.

Wagstaff, I., 1996, Prospects for Europe's Automotive Components Market, Economist Intelligence Unit, Londra.

Wilhelm, B., 1997, Platform and Modular Concepts at Volkswagen – Their Effects on the Assembly Process, in Shimokawa et al. (eds.), 1997.

Womack, J.P., Jones, D.T., Roos, D., 1990, The Machine That Changed the World, Macmillan, New York.

4 Competitive strategies, industrial models and assembly automation templates

Michel Freyssenet

For a long time, researchers and practitioners considered manufacturing automation as the result of scientific and technical progress and thus as a growth factor exogenous to the production sphere. The speed with which it was adopted and spread, together with its perfection could vary depending on financial means and available competencies; however, many thought they could not do without if they wanted to increase productivity and quality. On a different level, there was a debate on the organizational and social effects of automation. This debate can usefully be summarized into the confrontation of three main research orientations by over simplifying it. For some, productive automation entailed a change in the content and the organization of work, and consequently in the competencies needed (Touraine, 1955), for others, it had no effect on work itself, organization choices were the only important things, and their origin was to be found in the education system and professional relationships of a given country (Maurice, 1980). Although it was considered marginal for a long time, a third research orientation is making its way to the front line today: productive automation is grasped as an endogenous and multiple phenomenon. It would assume different forms depending on the objectives and the economic and social presupposition which direct its conception and implementation. It would then effectively have organisational and social effects, and these would vary according to the forms it has, because it is a social construction in itself (Noble 1989, Freyssenet, 1992, 1997). The present chapter will try to develop this type of analysis by widening this point of view. In some of my previous work, I tried to show how the image that managers and engineers have of productivity and man at work bears a consequence on their automation choices (Freyssenet, 1989, 1994). Here, I will demonstrate that these images are part of competitive strategies which differ from firm to firm. Two series of research will be the base for this demonstration.

The diversity in strategy and automation forms adopted by car manufacturers in the 80s and 90s is one of the main conclusions reached by Koichi Shimokawa, Ulrich Jurgens and Takahiro Fujimoto in their research work on assembly automation following two colloquiums they organised: Berlin in 1992 and Tokyo in 1993. As they point out in the book they co-wrote *Transforming Automobile Industry, Experience in Automation and Work Organisation,* this world-wide diversity can also be found in three regional poles for car production (Europe, North America, Japan), even if it is not always similar. Thus, strategies and automation forms are not linked to one particular company management model, be it national or regional, should such models exist.

So what are the causes for this diversity? According to Koichi Shimokawa, Ulrich Jurgens and Takahiro Fujimoto, the causes are the objectives, the contexts and the time of automation which have differed from firm to firm. Of course, everybody is concerned with competitiveness, they write. But some manufacturers brought automation to improve working conditions, others to face the variety in demand, others to increase the product quality and others yet to master new technologies. These differences in priority are rooted either in the constraints of the work market, or in the changes in car demand, or in the relationship with other manufacturers. As for the future, Koichi Shimokawa, Ulrich Jurgens and Takahiro Fujimoto believe that the next assembly automated systems will be the result of the hybrid of the various present strategies and assembly systems, and their making will depend on the history, the competence and the context of each firm.

I would like to go further in this analysis, starting not only from the cases reviewed in their book, but also from the results of the international research program of the GERPISA, carried out from 1993 to 1996 on "Emergence of New Industrial Models in the Car Industry ", that I co-ordinated with Robert Boyer (Freyssenet, Mair, Shimizu, Volpato, 1998; Boyer, Charron, Jurgens, Tolliday, 1998; Durand, Castillo, Stewart, 1998; Lung, Chanaron, Fujimoto, Raff, forthcoming; Boyer, Freyssenet, forthcoming) and from the work I carried out on automation, in particular in Renault (Freyssenet 1992, 1994, Decoster, Freyssenet, 1997). The object of the first section of this chapter is to spell out the questions left unresolved and which give an understanding of the diversity of strategies and automation forms, after a remainder of what Takahiro Fujimoto, co-author of the said book, said in particular with respect to characteristics and explanations. The second section will concentrate on understanding the basic reasons for diversity from the differences implemented by companies in competitive strategies and from the industrial models they have built to put into practice these strategies within their own context. The third section will show how the others causes for the diversity in strategies and in automation forms come from the difficulties in building or using an industrial model in order to put into practice efficiently the chosen competitive strategies.

4.1
The diversity of strategies and of automation forms. Unresolved questions

From the cases presented in the book and a very precise questionnaire survey carried out with the twelve Japanese car manufacturers, Takahiro Fujimoto identified four main automation strategies which differ in what the manufacturers want to improve and the means to reach that goal. What they want to improve can be spread along an axis from working conditions to the firm competitiveness. The means used can also be spread along an axis going from an additive approach of automation to a systemic approach. Positioned perpendicularly, these two axes are the limits of four quadrants, corresponding to the four identified strategies, each inferring a particular form of automation. We will summarise them briefly and invite the reader to go to the original text for more details (Fujimoto, 1997). The "Human Fitting Automation" strategy aims to improve working conditions by automating the most physically demanding jobs one by one. It was adopted in the Toyota Tahara factory, and also in the Nissan Kyushu one. The "Human-Motivating Automation" strategy aims not only to improve working conditions, but also the work content using automation in particular to allow one working team to completely assemble part or the whole of a vehicle. It could characterise the Volvo-Uddevalla factory, and also the Toyota Kyushu one in certain respects. In the "Low-Cost Automation" strategy, automation choices are dependent on the cost-cutting objective in all aspects, including investments. Automation is limited in its spread and level to what is economically relevant.. This happened in the Toyota factories in the 80s. The main features of the "High Tech Automation" strategy, which is based on a kind of technological optimism, are sophisticated pieces of equipment supposedly enabling one to reach a high level of productivity, quality and flexibility. According to the author, it was applied in Volkswagen Hall 54 in Volsburg and in the Fiat-Cassino factory.

This typology is a new and important contribution to the effort made to substitute to the classical representation of automation, the new vision of automation. The classical representation is homogeneous in its technical characteristics, changing with scientific and technical discoveries, and spreading out more or less quickly depending on the countries and the firms. The new vision of automation is varied in its objectives and forms, following the contexts which can be dominated either by work market requirements or by product market constraints, and because of approaches which can be systemic or on the contrary additive. Indeed, firms could not adopt the same strategies and the same forms of automation, as they had to solve different problems at different times and with different means.

The proposed typology enables us to understand why, in certain cases, the priority of automation is the improvement in working conditions and in others an increase in competitiveness, given the different economic and social contexts. On the other hand, it does not help us to understand why in the same context the additive approach is given preference over the systemic approach, or vice-versa. It does not help either to understand why (and the case is real) the improvement of working conditions was chosen in a context of competition and demand constraints.

Further to their main features, the forms of automation observed show secondary characteristics, the combination of which must also be explained. A first series of choices are in effect the general scheme of the assembly, that is to say: the type of assembling (complete assembling of vehicles in fixed and parallel stations as in Volvo-Uddevalla, additive line assembling as in nearly all assembling factories), the structure of the line (long, short, chopped, with secondary assembling lines for sub-units), the way the product moves (a non-stop or step-by-step conveyor, "Automated Guided Vehicles"). This first series of choices affects the following ones. The second series is the spread of automation, the types of jobs to be automated, the location of automated equipment in relation to the zones that remained manually handled (are they grouped in separate zones or are they scattered in the manually handled zones?) and in relation to how the product moves (is it fixed and sensitive, or mobile and synchronised with the line, or is it a autonomous transfer line in itself?). The third choice is the level of adaptability of the automatic tools themselves and the part left for the worker: is the equipment mono or multi-product (through adaptation, sub-division or derivation of the line, or AGVs)? Are the part alignments done mechanically, with manual guiding or automatically with a video camera? What is the part left for the worker?

GERPISA international program GERPISA "Emergence of New Industrial Models" allowed the study of the requirements of both work and the market, their origins, their long-term changes in the three world poles of car manufacturing, and the different means used by firms to answer them, thanks to the analysis of their trajectories et hybridisation processes. It would seem that many of the means used by firms are in fact part of the competitive strategies and the industrial models that allow for such strategies to be carried out. In the following section, we will study these strategies and models to see what they imply in terms of automation strategies and forms.

4.2
The firms' competitive strategies and the industrial models adopted to carry them out imply different forms of automation

A competitive strategy is characterised by the source(s) of profit it privileges: volume, diversity, reliability, the equipment level, innovation, productive flexibility, technical changes, cost reduction for constant volume. In order to implement it effectively, one must have a product policy, a productive organisation and a wage structure which are consistent with it, in a word, an industrial model. There may be various ways of implementing continuously the same competitive strategy. Hence, several industrial models can be built for one strategy only. It all depends on the management compromise that can be drawn up, bearing in mind the era and the environment, between the main actors: shareholders, managers, employees, trade unions, suppliers and distributors, public services.

The strategy making economies of scale a priority was efficiently put into practice through the Ford model. Today, no firm will produce a sole high volume model, like Ford did from 1909 to 1927 or Volkswagen from 1947 to 1973. However, some firms will produce three or four models only, each with its own specific platform, with only one or two body possibilities and very few options. This is a volume strategy only for the market segments that make it viable. It supposes that an important part of the customers of each of these segments has homogeneous needs and expectations on the long term. The Ford model is the one that historically implemented this strategy by specialising lines, men and machines for a number of years, and by constantly trying to increase productivity. As we know, this model is founded on a compromise characterised by employees accepting repetitive work and the constraints of productivity growth in exchange for a regular increase in the purchasing power of their salaries. It can only remain viable if this company management compromise lasts. When the market conditions are the ones mentioned above (as in South-Korea for about twelve years), automation comes as specialised equipment that can function for a while at a high pace, i.e. all at once stiff, simple to use and solid equipment (Chung 1998). Initial investments are high, but they are paid off through the large volumes. Equipment is only very partially reusable for the subsequent car models, but their maintenance is cheaper thanks to their lack of sophistication. For stamping, as well as for welding, automated specialised transfer lines are usually used. Because of the relative complexity of the product and of the manual operations that remain, it has been difficult to automate assembling. Joining of the body and mechanical parts, installing the wheels and the frontends are operations that can be automated on this mode. They constitute automated zones distinct from the manual zones. The part left for the worker remains limited. Manual operations become even more repetitive because of the reduction in the cycle length, the operating of the automated specialised transfer lines can be left to operators without any particular skills.

The second clearly identifiable competitive strategy is the strategy which puts first and makes the economies of scale and the effects of variety compatible. It consists in increasing the number of car models offered while commonizing the biggest possible number of their parts and pieces. It supposes a diversity in the expectations of customers towards the automobile following a relatively regular continuum from the bottom to the top-of-the-range and hence a low differentiated income distribution. Customers want a diversity in use, appearance and equipment but accept numerous non-visible common elements. In order to remain profitable, this strategy needs either that the demand for this type of product range increases both nationally and internationally, or that the firm widen its market share by buying out other firms keeping their models but installing on them common mechanical parts and components. It has been pursued by firms like General Motors since the 20s, Ford, Renault, Peugeot, Fiat, Nissan since the 60s and Volkswagen since 1974. The industrial model which successfully implemented this strategy is the Sloanist model, and only General Motors from 1945 to 1974 and Volkswagen since 1974 have succeeded to live on a long term basis (Freyssenet 1998). It is founded on a company compromise which consists in the workers to accept a repetitive and polyvalent job, in exchange for career and salary progression depending on the number of jobs. Assembling lines as well as machines and manufacturing workers must be able to ensure diversity and variation in the volume to be produced. The lines are chopped with buffer stocks allowing to absorb assembling time variation depending on the type of vehicle being assembled and various incidents due to diversity, and to prevent line stoppage. Automated machines are adaptable to the variety of bodies and equipment levels of cars. The specialised transfer lines enable the rapid replacement and tuning of press matrices. Welding lines are made up of a multi-product main line (thanks to robots or specialised welding machines put in derivation), fed by secondary lines or welding units. This fishbone structure tends to be more frequent in assembling. Indeed, it allows for diversity to be put upstream the main line, where automated equipment is concentrated with module fixation operations. The "wedding" of the body and the underbody is done in multi-products fixed automated installations (Wilhelm 1997, Naitoh, Yamamoto, Kodama, Honda 1998, Decoster, Freyssenet 1998, Jürgens 1998). The automated installations are sophisticated and compact. The operators do not have to control their functioning, but only to intervene to restart production fast in case of stoppage, leaving repairs to maintenance workers and reliability to technicians and engineers.

The innovation and flexibility competitive strategy consists in the conception of vehicles corresponding to emerging demands, and in their massive production if the orders confirm what was anticipated. This must be done before the other car manufacturers (those pursuing a 'volume and diversity' strategy in particular) come and copy them. It was the strategy chosen by firms like Chrysler, Honda, Mitsubishi, and more recently Renault. It supposes societies in which new social groups appear with their own expectations and needs. Income redistribution is advantageous for new competencies appearing or for initiatives that are taken, all this without creating considerable inequalities. Global demand can grow or not, but it renews itself within its structure. Thus the range of models is not organised in a regular hierarchy following a stable income scale with a foreseeable evolution, but it is organised in consistent models, with their own platform, according to the expectations of the emerging social groups. The Honda industrial model is the one that successfully represents this competitive strategy, as the other firms which pursued this strategy have never up until now been able to acquire consistent means to make up an industrial model. The Honda model is founded on a company management compromise where expertise and individual initiative are valued at all levels in exchange for the best working conditions and salaries in the sector (Mair, 1998). Above all, the socio-productive organisation allows the rapid allocation means and labour to the innovative vehicle(s) which has (have) found its (their) customers. It is characterised not only by ergonomic work stations but also by the fact that these can be adapted by the workers. The manufacturing integration rate is low. Assembling lines are short and easily adaptable. Parts are delivered by the suppliers assembled in small and medium-size sub-units. Automation in assembling is in fact reduced to assisting systems for the joining of heavy or bulky components. The important part left for the worker and the improvement of working conditions are inherent to the Honda model. These characteristics preceded the work market slump in Japan from 1988 to 1992, even if this led Honda to experience even more "human" organisations (Tanase, Matsuo, Shimokawa, 1998).

The fourth strategy is the one putting first the permanent reduction of costs for constant volume. Even when a market allows for example economies of scale, or to have quality paid for, the company carries on regardless to reduce its costs for constant volume, as one is never safe from a change in the economic set-up, the failure of a model, a change in the state economic policy, a variation in the exchange rate or the competition of a more effective manufacturer. In order to be successful, this strategy supposes to find means that make it acceptable to employees, suppliers, ... and even competitors. Indeed it is very demanding for the employees and the suppliers and exacting for competitors. Toyota succeeded in doing this for about 40 years (1952-1990), obtaining a direct contribution from its employees and suppliers in the reduction of standard time slots (Shimizu, K., 1998). They managed to limit their competitive advantage to the point beyond which protectionist reactions would have prevented any international expansion. The company management compromise founding this model is the participation and permanent reduction of costs in exchange for employment and salary increase guarantees for employees and order guarantee for suppliers. In general, reorganisation is used first for the improvement of performances, the introduction of new machines comes after. Existing machines are improved and transformed to their maximum before they are replaced. These organisational and technical changes prepare for automation, which then happens progressively and at the lowest costs. The lines are continuous without buffer stocks, so that problems of line stoppage appear immediately and are solved rapidly. For the same reason, automated equipment has a low level of sophistication, that way the causes of breakdown and anomalies are easily found. The level of automation is only increased following the knowledge and control of perturbing factors. These are spread throughout the assembling lines. The operators of nearby manual work stations intervene as whenever it is necessary.

The fifth competitive strategy is the one putting first specialisation in top-of-the-range and the quality of vehicles. It is (was, for some maybe) the strategy of BMW, Mercedes, Saab, Volvo. It supposes that part of the customers are ready to pay the price for quality and social difference, i.e. well off customers. The manufacturer who adopts this strategy is not looking for volume. Too wide a diffusion would make the product ordinary and reduce the attraction. Margins are made on a high price. This price is part of the product definition as much as the quality of the mechanics, the equipment and the finishing and the other socially distinctive signs. The customers concerned are more or less spread out depending on the type of redistribution of the country national income, but by nature remains limited. That is why manufacturers choosing this strategy aim both at the domestic and international markets right from the start. The management compromise of these companies is based on the diversity and quality of work from the workers in exchange for better working, salary and stability conditions than with other more general manufacturers. The workforce is more polyvalent and used to option and version variations for the same model. Cycle times are longer as speed is less important than execution quality. Neither the press line nor the welding line are high paced. They do not need to be highly automated. Assembling lines are chopped and separated with large buffer stocks. There are numerous help systems for joining operations, as well as ergonomic tools making the work easier, such as the rolling-axis transport system (Hsieh, Schmahls, Seliger 1997). The line can be replaced by AGVs which, though expensive, give even more flexibility and above all better working conditions. Having the specialised German and Swedish manufacturers looking at automation from the point of view of the improvement of working conditions is not that surprising, even though they were faced with the same constraints from the work market as major European firms. The lengthening of the cycle time, work groups, assembling in fixed modules, joining assistance, and transfer automation also enabled a response to increased requirements in quality and to manage diversity more easily. These are essential conditions for the specialised manufacturers. The experiences went the furthest with them, all the way to the complete suppression of the assembling line principle and its replacement with complete assembling of cars in fixed parallel stations with two or four workers (Engström, Medbo, 1995, Ellegard, 1995, 1997, Nilsson, 1995).

The sixth strategy is the variety and flexibility one. It characterised British manufacturers in the 50s and 60s, and Mazda in the 80s, however this manufacturer was unable to use it successfully due to an inappropriate market for this strategy and to insufficiently consistent means. Indeed, the "variety and flexibility" strategy consists in offering car models to social groups with specific needs and expectations. It supposes a society where the distribution of national income is all at once unequal and stratified. It is quite possible that the conditions necessary for such a strategy to be successful might reappear with the deregulation of salary scales and employment structures. This strategy requires the rapid creation of new models and production in medium run of a wide variety of vehicles with few parts in common. That is why vehicle modularization, manual assembling of modules on sub-assembling lines and automated fixing modules on the main line are a way towards solving the difficulty (Kinutami 1997).

The last competitive strategy is the production of luxury and sports cars pursued by Rolls-Royce, Porsche, Maserati, Lamborghini, etc. It consists in creating and producing luxury and sport cars in very short runs and for a very high price, symbolising all at once the wealth and personal tastes of the owner through the level of personalisation, notably in the equipment. The production of a manufacturing type in fixed stations does not prevent automation, in particular assistance from machines with digital command for the machining and the fitting of parts.

4.3
It is difficult to find consistence between the competitive strategy, the socio-productive organisation and the form of automation

If firms applied the automation form consistent with the industrial model they created or use, the number of forms of automation would be limited to the number of identified industrial models. Reality is of course much more diverse. Not only do firms sharing the same context and pursuing the same competitive strategy turn to different forms of automation, but they also implement different forms of automation within their plants. Does this mean there is no rule? The diversity inter and intra-firm, larger than the number of existing industrial models, can be explained by two main facts: firms embodying a model must change from time to time, firms that cannot find a management compromise to bring consistence between the various means used are by far the most numerous. Hence multiple and sometimes contradictory automation forms.

All the firms that either created or used one particular industrial model met with difficulties after a more or less long period of success, and have had to change their socio-productive organisation, their employee relations, or even their product policy. There are two reasons to this: either success modified the conditions that made the company management compromise possible; or, the type of growth and redistribution of national income changed, following changes in the competitive relationships between countries, and thus the competitive strategy pursued lost its pertinence (Boyer, Freyssenet, forthcoming). General Motors met with these two types situations one after the other. They implemented the "volume and diversity" competitive strategy successfully inventing and embodying the Sloanian model in the 50s and 60s, to the point that many firms tried to imitate it at the time (Freyssenet, 1998). But the entering of a replacement market at the end of the 60s, the difficulty in exporting American cars because of their specificities, and the impossibility to absorb another manufacturer stopped the volume growth indispensable to this strategy, as well as the increase in the staff, the internal professional mobility and the salary progression which are all part of the Sloanist management compromise. The dynamic of the model was further stopped by the 1974 slump. Due to the increase of the energy bill and of the competition, the USA has had to change the way national income was redistributed to become competitive. With escalating social and economic inequalities, the "volume and diversity" strategy lost in pertinence. GM had to add to the variety of their offer without being able to increase the communising rate of their platform. With these difficulties, GM first tried to improve their productivity and the quality of their products by giving way early, beginning of the 70s, to automation and robotization, while the origin of their problems were to be found in products ill adapted to exports and which were beginning to become so on the domestic market. Automation was GM trying to adapt, thinking they could keep their model and strategy. Actually, not only was automation inappropriate and costly to solve GM's competitiveness, but it also enhanced it bringing about further staff reductions and further destabilisation of the company management compromise.

In 1990, the Toyota model was presented as the example of what the future industrial model should be. The same year, this very model reached the limit of social acceptance within Toyota. The boom in Japan domestic demand and the impossibility to recruit enough young employees because of harsh working conditions led supervisors and workers to refuse the increase in overtime, and on a larger scale the work and salary system. Toyota then had to overturn their production system and to find new ways of to involve employees in cost reduction for constant volume (Shimizu, 1998). They drastically reduced, or even suppressed as in the Toyota-Kyushu factory, the obligation that each team had to reduce their standard assembling times. They chopped assembling lines in separate parts with buffer stocks to limit the pressure from continuous flow. They rearranged work stations to suppress painful positions. The automation form was affected by this. Automated equipment was simplified, its functioning made even more visible and controllable to the workers of the team, so that they could improve it if necessary (Niimi, Matsudaira, 1997). But this new form coexists with ancient forms that will only disappear with a change in the car models. It is also possible that the new situation of the work market in Japan and the changes in Toyota product policy will lead to yet another form before the present one spreads out. Indeed, Toyota is searching. They have not managed to find again the internal consistence they had for several decades. It is also possible that they are changing their competitive strategy, as certain new directions and manager announcements would lead us to believe.

But the diversity in automation forms is mostly due to most firms' incapability to have some consistence between product policy, their productive organisation and their employee relations and with the competitive strategy they are pursuing. The reasons are many.

Not many realise the necessity of consistence, in particular of technical choices with social choices and strategic choices. For them, these choices are in different domains with their own independent logic. That is why one can see many cases where automated installations are compact for space saving and difficult to reach for reasons of security, whereas it is officially required of operators-supervisors to closely follow the functioning of these installations to prevent or see anomalies, so that they can find and take care of the primary causes, and thus narrow the flow and reduce the assembling time (Freyssenet, 1994).

The construction of a long-term company management compromise around the means to be used is a very difficult process to bring to its term. Beyond the fact that some of the actors are not necessarily conscious of the importance of this compromise for the life of the company, they have to be able to create it, i.e. to have at the same time macro-economic and society conditions which hardly ever happens as they are the result of quite unintentional processes. A number of employees and trade unions in Europe, the USA and also Japan thought for a long time that it was up to car manufacturers to improve continuously the employee situations, and that the managers should find the means alone. Some of the latter were not far from the idea themselves. The imperative of competitiveness following the oil crisis reminded the main actors of their necessary implication in the construction of a new management compromise. But the rapid increase of unemployment in many countries, the softening of employment laws and the weakening of trade unions created a political and social situation that allowed many firms to do without the real construction of this new compromise, with employees being obliged to accept the new working norms. Since many things had become possible, firms were deprived of the obligation of consistence imposed by any one management compromise. Hence some of the easy, immediate and one-off earnings that they get are sometimes lost due to the malfunctioning they generate.

Competitive strategy and management compromise are the subject of debates within the firm, when the different parties admit their necessity. Groups argue about their pertinence and their lasting. One method of influencing one way or the other is to have decisions made on aspects of the organisation that bear neither any apparent link with the firm's strategy nor with the internal balance between the actors. The sheer accumulation of these decisions will make one particular orientation irrevocable. This is the way with automation forms which can be justified with technical considerations, or with the immediate earnings in personnel that they bring, or still with the improvement in working conditions that they supposedly make, whereas they carry a new orientation of the competitive strategy, because of the financial means they take from other uses, or a hidden change in the socio-productive organisation.

The automation form to be used is not directly, easily and precisely deducted from the competitive strategy pursued, nor from the construction of the management compromise. What rate of flexibility should one try to reach for automated equipment when one is pursuing a "volume and diversity" strategy and looking for employees participation, such as Fiat, for example? The way this manufacturer hesitated when trying to find the adequate automation form, from the "Mascherone automatico" to the "Fabbrica Integrata" having used the "Robogate", clearly illustrates the difficulty (Camuffo and Volpato, 1997).

Leading a company is a difficult task, and often enough, managers who cannot find the specific solutions resort to imitate what seems to be working with their competitors. To start with, this process leads us to believe in a convergence towards similar solutions. But the different results obtained with the same technical solution come to show once more that these results depend on the conditions in which a technique is used as much as the technique itself. A solution can only be the answer, i.e. contributes to the upturn of the company's performances, if it is consistent with the rest of the organisation and the strategy.

Lastly, company managers do not always realise which competitive strategy they are pursuing, or more precisely which is effectively at the start of their performance. As a consequence, they might be led to inconsistent automation choices. Other managers believe they can conciliate several competitive strategies. This is how, in the recent past, several manufacturers have been offering to create and produce innovative models as an answer to the new expectations of one part of the customers, as they appear in a number of countries, and this on top of their classical offer. But unlike a new model within the classical offer, an innovative model has one particularity: either it is a failure, or its success comes as a surprise. This means a flexible company in all areas: financial, social, organisational and technical, one which can failure or success. And a firm pursuing a classical product policy is not structured to that effect.

4.4
Conclusion

The diversity in automation forms between the different firms lies first in their difference in competitive strategy and industrial models. Secondly, it lies in the inescapable crisis met by these models, often then generating multiple and contradictory technical forms within the firms themselves. Last but not least, it lies in the difficulty for firms to embody completely one particular industrial model. This would mean that they have to create and keep alive a company management compromise between the main actors on the means to be used (product policy, productive organisation, employment relationships) to implement the competitive strategy pursued, the economic and social pertinence of which might be rudely called into question because of a change in the type of growth and national income redistribution, as has happened twice since the 60s.

4.5
References

Boyer, R., Charron, E., Jürgens, U., Tolliday, S., eds (1998), Between Imitation and Innovation: The Transfer and Hybridization of Productive Models in the International Automobile Industry GERPISA books, Oxford University Press, Oxford, 376 p.

Boyer, R., Freyssenet, M. (forthcoming), The World that Changed the Machine , GERPISA books.

Camuffo, A., Volpato, G. (1997), "Building Capabilities in Assembly Automation: Fiat's Experiences from Robogate to the Melfi Plant" in Shimokawa, K., Jurgens, U. and Fujimoto, T. (eds), Transforming Automobile Assembly. Experiences in Automation and Work Organization Springer, Berlin, pp.167-188.

Chung, M-K. (1998), "Hyundai Tries Two Industrial Models to Penetrate Global Markets", in Freyssenet, M., Mair, A., Shimizu, K., Volpato, G., eds (1998), One Best Way? Trajectories and Industrial Models of the World's Automobile Producers , GERPISA books, Oxford University Press, Oxford, pp. 154- 175.

Decoster, F., Freyssenet, M. (1997), "Automation at Renault: Strategy and Form" in Shimokawa, K., Jurgens, U. and Fujimoto, T. (eds), Transforming Automobile Assembly. Experiences in Automation and Work Organization , Springer, Berlin, pp. 157-166.

Durand, J.D., Castillo, J.J., Stewart, P., eds (1998), "Teamwork in the Automobile Industry: Radical Change or Passing Fashion?" GERPISA books, MacMillan, Basingstocke, 368 p.

Ellegard, K. (1995), "The Creation of a New Production System at the Volvo Automobile Assembly Plant in Uddevalla, Sweden" in Sandberg, A. (ed), Enriching Production, Avebury, Aldershot, pp 37-60

Ellegard, K. (1997), "Worker-Generated Production Improvements in a Reflective Production System" in Shimokawa, K., Jurgens, U. and Fujimoto, T. (eds), Transforming Automobile Assembly. Experiences in Automation and Work Organization, Springer, Berlin, pp. 318-334.

Engström, T., Medbo, L., (1995), "Production System Design, a Brief Summary of Some Swedish Design Effort" in Sandberg, A. (ed), Enriching Production, Avebury, Aldershot, pp 61-74.

Freyssenet, M. (1992) "Processus et formes sociales d'automatisation. Le paradigme sociologique" Sociologie du Travail, vol 2, pp. 469-495.

Freyssenet, M. (1994), avec la collaboration de Charron, E. et Beaujeu, F., L'automatisation du montage automobile. Conceptions techniques, organisationnelles, gestionnaires et sociales. Divergences et conditions de mise en cohérence , CSU, Paris, 132 p.

Freyssenet, M. (1997) "The Current Social Form of Automation and a Conceivable Alternative: Experience in France" in Shimokawa, K., Jurgens, U. and Fujimoto, T. (eds), Transforming Automobile Assembly. Experiences in Automation and Work Organization, Springer, Berlin, pp. 305-17.

Freyssenet, M., Mair, A., Shimizu, K., Volpato, G., eds (1998), One Best Way? Trajectories and Industrial Models of the World's Automobile Producers , GERPISA books, Oxford University Press, Oxford, 475 p.

Freyssenet, M. (1998), "Intersecting Trajectories and Model Changes" in Freyssenet, M., Mair, A., Shimizu, K., Volpato, G., eds, (1998), One Best Way? Trajectories and Industrial Models of the World's Automobile Producers , GERPISA books, Oxford University Press, Oxford, pp. 8-48.

Hsieh, L-H., Schmahls, T. , Seliger, G. (1997), "Assembly Automation in Europe. Past Experience and Future Trends" in Shimokawa, K., Jurgens, U. and Fujimoto, T. (eds), Transforming Automobile Assembly. Experiences in Automation and Work Organization , Springer, Berlin, pp.19-37.

Jürgens, U. (1997), "Rolling Back Cycle Times: The Renaissanceof the Classic Assembly Line in Final Assembly", in Shimokawa, K., Jurgens, U. and Fujimoto, T. (eds), Transforming Automobile Assembly. Experiences in Automation and Work Organization Springer, Berlin, pp. 255-273.

Kinutani, H. (1997), "Modular Assembly in Mixed-Model Production at Mazda" in Shimokawa, K., Jurgens, U. and Fujimoto, T. (eds_Transforming Automobile Assembly. Experiences in Automation and Work Organization , Springer, Berlin, pp.94-108.

Lung, Y., Chanaron, J-J., Fujimoto, T., Raff, D. (forthcoming), Coping with Variety: Product Variety and Production Organization in the World Automobile Industry , GERPISA books.

Mair, A. (1998), "The Globalization of Honda's Product-Led Mass Production System" in Freyssenet, M., Mair, A., Shimizu, K., Volpato, G, eds. (1998) One Best Way? Trajectories and Industrial Models of the World's Automobile Producers , GERPISA books, Oxford University Press, Oxford, pp110-128.

Maurice, M. (1980), "Le déterminisme technologique dans la sociologie du travail, 1955-1980. Un changement de paradigme", Sociologie du Travail, n∫1.

Naitoh, T., Yamamoto,K., Kodama, Y., Honda, S. (1997), "The Development of an Intelligent Body Assembly System" in Shimokawa, K., Jurgens, U. and Fujimoto, T. (eds) Transforming Automobile Assembly. Experiences in Automation and Work Organization , Springer, Berlin, pp. 121-132.

Niimi, A., Matsudaira, Y., (1997), "Development of a New Vehicle Assembly Line at Toyota: Worker-Oriented, Autonomous, New assembly System" in Shimokawa, K., Jurgens, U. and Fujimoto, T. (eds), Transforming Automobile Assembly. Experiences in Automation and Work Organization Springer, Berlin, pp. 82-93.

Nilsson, L., (1995), "The Uddevalla Plant: Why did it Succeed with a Holistic Approach and Why did it Come to an End?", in Sandberg, A. (ed), Enriching Production, Avebury, Aldershot, pp 75-86.

Noble, D. (1989), Forces of Production. A Social History of Industrial Automation, Alfred Knopf, New York.

Takahiro, F. (1997), "Strategies for Assembly Automation in the Automobile Industry" in Shimokawa, K., Jurgens, U. and Fujimoto, T. (eds) Transforming Automobile Assembly. Experiences in Automation and Work Organization, Springer, Berlin, pp. 211-37.

Tanase, K., Matsuo, T., Shimokawa, K. (1997), "Production of the NSX at Honda: an Alternative Direction for Assembly Organization" in Shimokawa, K., Jurgens, U. and Fujimoto, T. (eds) Transforming Automobile Assembly. Experiences in Automation and Work Organization, Springer, Berlin, pp.109-120.

Touraine, A. (1955), L'évolution du travail ouvrier aux Usines Renault , CNRS, Paris.

Shimizu, K. (1998), "A New Toyotaism?" in Freyssenet, M., Mair, A., Shimizu, K., Volpato, G, eds. (1998), One Best Way? Trajectories and Industrial Models of the World's Automobile Producers GERPISA books, Oxford University Press, Oxford, pp. 63-90.

Shimokawa, K., Jurgens, U. and Fujimoto, T. eds (1997), Transforming Automobile Assembly. Experiences in Automation and Work Organization , Springer, Berlin, 414 p.

Wilhelm, B. (1997) "Platform and Modular Concepts of Volkswagen. Their effects on the Assembly Process" in Shimokawa, K., Jurgens, U. and Fujimoto, T. (eds), Transforming Automobile Assembly. Experiences in Automation and Work Organization , Springer, Berlin, pp. 146-156.

5   Automation and Inertia

Kajsa Ellegård

5.1
Change and profitability

The aspiration for profitability is the fundamental driving force behind changes in industry, and many different measures are taken to reach profits. Companies steadily strive for improved product quality in order to satisfy the wants and needs of customers, thereby hoping for increased market shares. In many industrial production processes it is profitable to make investments in new techniques that make production more rational in terms of increased flexibility and higher efficiency. Therefore, many arguments for industrial change are based on technical aspects.

At times, however, managers use social arguments for industrial change. In Sweden socially based arguments were used during the 1970's and 1980's, emanating from work force related problems of that time, such as high costs for recruitment, high sick leave and high labour turn over rates etc. These problems in turn, were generated by low job satisfaction and, consequently, low commitment to work for the employer.

Both technical and social arguments have been used in favour of automation. On one hand, technical means are primarily used to solve technical problems, but technical change strongly affects the social structure of an organization, a fact that is not sufficiently taken into account in investment planning. On the other hand, social means are mostly used to solve social problems only. However, during the last decades a process has started in Swedish industries, wherein technical and social arguments in combination are used for investments, so that work organization is consciously changed when production is automated.

Automation, i.e. development of new techniques and implementing them into factory life has been a useful means for improving productivity and efficiency in industry. However, problems in the automation process often arise, and it does not matter whether the arguments for change primarily is of technical or social nature. Whenever social factors are not seriously taken into account and when they are not amalgamated into the technical side of the implementation of a technique, one major problem arise, namely that the pace of change often is much slower than intended. The aim of my paper is to penetrate the inertia in processes of change by reflecting over a lengthy automation process in a car body shop. I will start with presenting a model for change that pinpoints the need for social embedding of automation processes.

5.2
A model for change

Even though great technical innovations have shown to be important means to increase profitability, the individual biographies built upon former experiences of the actors involved influence the process of change in a very powerful way. Thus, when it comes to the social management of technical change one handling strategy for managers emanates from their assumptions about what the actors involved know from experiences in the past. This is a convenient a convenient strategy as many things then can be taken for granted. If, for example, a new machine may be approached by middle managers and workers according to their old way of acting and being in the organization, even a huge technical change may be successfully implemented without big planned social changes, as it will work satisfactory anyway. In such a situation, decision makers and other people implementing the technical novelty make the process of change smooth because the social consequences of the change are *perceived* as limited, as people's conceptions, the organizational rules and its structure may remain more or less the same as before.

Generally my model of change leans upon the assumption that people orient themselves to a new technical situation as good as they can by means of milestones in their experiences from their own biographies. If this biography bound strategy of managing changes is not taken into consideration, the full potential of the new technique cannot be utilized. Hence, in industry, the past is embodied not only in the existing plants, but also in the minds of the managers and employees, and it is very difficult for managers to introduce radical social changes, especially in a traditional, hierarchical organization, where the employees' conceptions of production routines are fixed and cemented.

There are some situations, though, when big changes, social and technical, may be more easily introduced: First, if either societal or internal factors put the firm in a very threatful situation, a firm may be forced to make big changes. Second, if the company is very profitable and has a lot of money to spend for investments and for developing a new profile of innovative management.[1] Third, when a regular investment is about to be realized, either to replace old equipment or to increase capacity. I will concentrate on the third situation concerning regular investments. Such regular investments emphasize the physical and mental power of past experiences. To my understanding, a regular investment implemented in an existing industrial plant will be strongly influenced by what there already is in terms of techniques (machines and other equipment) and social agreements (formal and tacit rules and conceptions).

The process of change can be illustrated by a model, which is inspired by the school of time-geography within the human geography discipline. The model presented here has three dimensions, see fig. 1.

1. A technical dimension, with tools and hand machines located at one extreme point of the dimension, and automation at the other.
2. A social dimension, with unskilled labour performing repetitive work tasks organized in short cycles located at one extreme point of the dimension, and skilled labour used to problem solving, who are able to control their own work and work in long work cycles, at the other extreme point.
3. A time-dimension, where past, present and future may be located at a continual dimension.

In fig. 1 what has happened in the *past* is described by a trajectory (p to a) illustrating the movements along the three dimensions made by an organization. The trajectory in this example illustrates a movement towards deskilling (on the social dimension) and towards automation (on the technical dimension).

The *present* situation is described by the point (a), which illustrates the situation just now on the time dimension[2] and point (a) simultaneously locates the situation at the relevant point on the social and technical dimensions of change respectively. If we limit our exercise to solely two dimensions, focusing the technical and social dimensions and neglecting the time dimension, the present situation is illustrated by point (x) .

Finally, the *future* may be described by the aggregate of alternatives there are to chose among as they are perceived by the individual actors in the organization. The present situation is taken as the point of departure in order to find out what future alternatives one individual actor perceives. The individual may formulate more radical goals as the time horizon is widened towards the future, and consequently the number of possibilities perceived by the individual increases in the future. The smaller ellipse in fig. 1, thus, illustrates an 'individually perceived possibility space' of an individual actor that do not think of more radical

---

[1] This was the case when Volvo developed the reflective production system in   the planning of the Uddevalla plant and sucessfully put it into production. The
        reflective produciton system is a radical alternative to the traditional assembly
        line system of automobile industry.
[2] The past and the future, consequently, are identified by locating the present
        (now).

deviations from what there is now (at a). If we take all the individually perceived possibility spaces into account simultaneously an 'aggregate space of possibilities' (the big ellipse in fig. 1) may be defined. Each single individual's perceived possibility space usually is smaller than the 'aggregate space of possibilities' of the organization as a whole.[3] It is, however, possible to actively increase the size of the 'perceived possibility space' of individual actors in an organization. This is the big challenge for industrialists who want to be competitive in the future. The core question is: How can many 'individually perceived spaces of possibilities' in an existing organization be widened and how can this widening process be directed so that the 'aggregate space of possibilities' of the organization coincide with the goals set up for of the organization as a whole?

Within the aggregate space of possibilities' some outcomes are more probable to happen than others. The most probable outcomes of a process of a planned change in an organization are located in the area that easily are perceived by the majority of individual actors, an area which appeal to the individuals' experiences and knowledge about formal and informal rules of action. A common consequence of this situation is that in the future there will just be small deviations from what there is now, if no special arrangements are made.

---

3   In a hierachical organization the higher position an individual occupies, the
    bigger is his (her) 'individually perceived possibility space'.

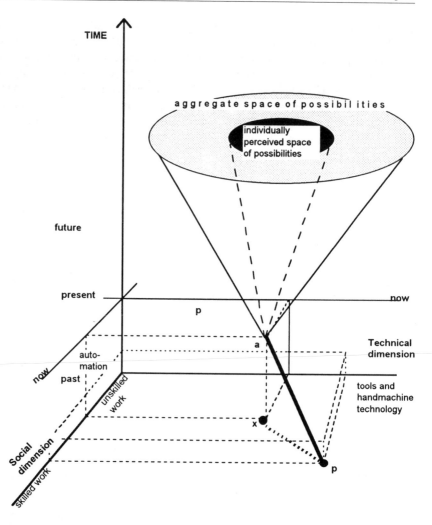

Figure 5.1 The dimensions of the model of change

The trajectory **p-a** describes the history of an organization moving from a position where work claimed for rather high skilled workers to a position where less skilled workers are needed (on the social dimension) in a plant where automation increased (on the technical dimension).

Note: The illustration aims at showing the *principles of the dimensions of the model of change*, it is *not* intended to illustrate the outcome of the process of automation in general.

5.3
Automation in industry

Blauner and Braverman wrote some decades ago about effects of automation from different experiences and perspectives. Blauner found that automation implies positive results for workers:

What emerges is a picture of automation made up of positive components, which are necessarily de-emphasized by those who concentrated on its immediate consequences for employment. The worker in the automated factory 'regains' a sense of control over his complex technological environment that is usually absent in mass production factories. (Blauner 1964).

Blauner's writings in the book 'Alienation and freedom' were oriented to the heavy and chemical industry (Blauner 1964). He found, of course, that the number of workers were reduced by automation in the process, but at the same time that the work content and work environment for the workers still employed was improved, as the demanding processing itself was embedded in the machines. The automation process created new work tasks, concerning process supervision and maintenance. These work tasks claimed for high skills.

Ten years later, Braverman wrote the book *Labour and monopoly capital* (Braverman 1974). He had a different type of work in mind from that of Blauner, namely his experiences from an industrial branch (printing works) where the craft traditions, and the vocational pride, was strong. Mechanization and automation in printing industry resulted in deskilling of the work force and, thus, that the interesting work tasks were taken away from the workers. Following Braverman, automation leads to a smaller, and in addition less skilled work force. Unskilled workers may serve the machines and perform residual work tasks that are either too expensive or (yet) impossible to automate. The workforce thereby lack work satisfaction, and the alienation among workers increases. The conclusion may be drawn that Braverman and Blauner are exponents of opposite explanations to what happens to work content when industry is transformed by automation.

Braverman and Blauner, thus, perceived different future possibilities as they drew conclusions from their very different experiences from different kinds of industry. Blauner presented a U-shaped curve, where he related the mood of production to a dimension showing degree of alienation. On the one axis he located craft production, mass production and heavy and chemical industry and on the other axis he located high to low degree of alienation.

Craft production was correlated to low alienation, mass production with its characteristic short cycle, tempo work showed a high degree of alienation and finally the automated heavy and chemical industries showed low alienation and higher skills. Braverman, on the contrary, concluded that automation leads to degradation of work. His conclusion was footed on the overwhelming impression of dequalification of individual workers, and the observation that the remaining manual work obviously was organized as residual tasks, complementary to what the machine manages. The conclusion of Braverman fits well into the concept of Blauner's theory as long as it comes to automation in craft production transformed into mass production industry. Thus, Bravermans work is of more limited scope than that of Blauner. Blauners theory may be used for describing differences between various kinds of industry. However, if Blauner's theory is extended with an explicit time dimension, it might in addition be used to identify phases in the automation process.

At a general level, however, the conclusions of Blauner and Braverman coincide: they drew the same general conclusion that employment in industry will decrease as a result of the process of automation. This conclusion holds true for the process of automation, but it is not the focus of this paper.

5.4
Inertia of an automation process, an example from car body manufacturing

5.4.1
Background

In this section I will give an example of an automation process by showing the development of automation in the bodyshop of Volvo Torslanda, Sweden. My first research project in Volvo started in 1979. My aim was to investigate the long term development of work content in car body production within Volvo from 1920's to early 1980's. In 1980, I made several interviews with workers on the shop floor in the body shop in Torslanda, and I also interviewed people in the plant where Volvo car bodies were manufactured until the early 1960's. It was clear that from the 1950's and till the late 1970's there were jobs at one work station in the body production process which had a relatively high status. At this station, called the main jig, the most skilled workers of the shop welded the bodyframe as they put the floor, sides, roof together to one unit. Their work required welding skills, strength, fitness and quickness. The workers on the main jig were the heroes of the shop floor, and young workers wanted to get there.

During the 1970's an intensive process of automation started in the body shop in the Torslanda plant. The first robot was installed there in the early 1970's. I will illustrate the inertia of this automation process by giving concrete examples and relate the entire process to the model of change, presented above. The automation process illustrated here is divided into three phases (first: technical changes, second: social changes and third: technical and social changes coincide simultaneously).

5.4.2
First phase of automation: technical changes

This first robot installation in the early 1970's aimed at testing a robot for spot welding. This robot was installed at one isolated station in a pre-assembly area. The main argument for robotization was technical rationalization and thereby a more cost effective production. This robot alone made the same welding as two workers previously did. No social measures were taken in order to improve work content for the workers, who now had to do the loading work at the station. Their new role was to serve the robot. There were no social argumentation for this first robot. When I in retrospect asked for social aspects people told me that heavy and dirty work was automated. The result for the workers, however, was that the monotonous loading work remained while the more demanding welding was automated. Job satisfaction decreased as the work tasks now became more monotonous.

The test of this first robot in the production yielded positive technical results, i. e. the robot worked with high precision, high quality, less people required for work and it was as reliable as could be expected. More technical tests were made to try other kinds of robots.

In 1975 a great step was taken when a number of robots in a line were installed. Six robots made the welding previously done by eleven workers. On this welding line, robots performed the final spot welding after the main jig, where the floor, sides, roof and window frames still were welded together manually. The technical arguments remained unchanged for automation, and there were still no other social arguments but to reduce the heavy and dirty spot welding jobs.

About one year later, the automation process took another big step, now entering the heart of the workers vocational pride in the body shop: the main jig with its high status jobs. Now all spot welding previously performed in the main jig were automated. Thereby overnight, all work tasks claiming for manual welding skills at the main jig were automated, and the work content was reduced to loading the machine and to make adjustments if there occurred small irregularities. Some spots were needed to hold the keep components in their correct geometric position, and they were welded by an automatic welding machine.[4] In the subsequent automatic welding line 27 robots were installed. Automation reduced the number of workers with fifty-eight. The remaining

---

4   An automatic welding machine can not move itself in more than two dimen
    sion, while a robot can move in three dimensions.

manual work was just loading activities. However, there were no social improvements made to compensate workers for the degraded work content. The cycle time had become half of what it was before (now about 80 seconds). The least negative effect seen from workers point of view was the technical buffer installed between the manual part of the process and the subsequent automated robot welding process. The (unexpected) positive social effect of this technically motivated buffer was that the workers were not completely bound to the pace of the robots. The buffer, however, was not installed for social reasons but for technical reasons in the case of machine break-down.

5.4.3
Second phase of automation: limited social improvements

In order to reduce the negative social effects of the automation, one measure of a social kind was taken after a period, when a job rotation scheme was introduced in the work stations at the beginning of the robot welding line. This meant that the seven workers performing the manual loading work tasks on the three stations in the beginning of the robot line, changed work tasks with each other after each break. However, the type of work, the cycle time and the level of skills requested, all remained the same. Rotation, though, made the workers repeat one and the same movement of their bodies less frequently each day, and thereby the ergonomic situation was improved in this respect. Still the level of skills required for workers was reduced.

A conclusion is that the robotline with its 27 robots was a first large scale try out of the relatively new robot technique, and until it was proven technically successful (reliability, quality, efficiency etc.) no in depth social improvements were made or even thought of. The workers on this robot line – skilled former main jig workers – all said that they felt their work degraded by the automation process.

So far, the automation process followed the path predicted by Braverman, with degraded, monotonous work tasks for the production workers.

5.4.4
Third phase of automation: technical development and social improvements

In 1980, it was obvious that the robotline technically was successful, but also that the workers were not satisfied with their work content. Many good production workers left the shop floor and some of them left Volvo. All workers who had an opportunity to get another job, for example in the department of quality, materials handling or maintenance, immediately took the chance to escape from production. Production was, by most people, looked upon as a place to flee from, as there seemed to be no possibilities to develop oneself, on the contrary: the work content was continuously degraded.

Thus, when a new body shop was about to be built in Torslanda in early 1980's, in order to increase capacity for a new car model, the unions and the managers of the body shop agreed upon a project to improve working conditions and to upgrade industrial work. The level of automation was about the same as in the old shop, but the number of robots increased. The project group was encouraged by the top management of the Volvo Group and the CEO PG Gyllenhammar to change the work organization as the labour turn over rates and sick leave figures were very high. Hence, there were social arguments for a social change. The intention was to turn the evaluation of work in the production department upside-down: the production department should, hereafter, be looked upon as the basic and most important ground for all activities in the body shop[5] and all other departments (quality control, adjustments, maintenance, materials handling, production engineering etc.) were to be looked upon as services to production, not as something superior to production. In the new organization some of the routine activities were transferred from service departments to the production department, and these activities were integrated into the work tasks of production workers, thus enriching their work content. This was a new way of thinking that, however, was not developed until the initial problems of the new robot technique as such was under control, which was done during the second phase of automation.

The union representatives and managers in the project group made an agreement on a new work organization in the early 1980's. This meant that a great social change and a minor technical change were made simultaneously. Unfortunately, though, the new work organization was not taken seriously enough by some of the first line and middle managers and engineers, and therefore it was not fully implemented until about ten years later. The middle managers of the early 1980's did not know how to handle such an organized social change, as changes they experienced before solely had been technically oriented, and their role then mainly had been to correct workers performance. Middle managers were now exposed to the power of their own experiences and thus they went on performing as usual, i.e. waiting for worker related problems to appear so that they could take measures against it. One reason for this inertia is that the fundamental difference between implementing a new technique and a new social organization respectively, even though the implementations are made in one and the same shop: New technical equipment, on one hand, is a tangible thing that proves its existence by itself, and by its mere existence it communicates to the worker that he has to perform his work task in a new way. The work can

---

5   This is called the core-activities in modern terminology.

not be done in the same way as before, simply for material reasons. Therefore the managers have little to do with the technical implementation as such, but they may have a lot to do with the correction of workers' performance.

New social organization, on the other hand, exists solely through the acts of people involved. The most tangible part of the new social organization is the piece of paper on which the words of the agreement is written down. If this piece of paper stays on the desk of a middle manager, or if he does not change the every day routines of his department and educates his employees according to the new routines, nothing will change. The power of what has been before, thus, is strongly rooted in the old social organization due to its lacking materialization. One conclusion from this lengthy implementation process is that first line and middle managers need a lot of education and training in handling social changes. Another conclusion is that if the profit potential of a technical change is to be realized, deliberate measures are to be taken to implement it not only in the physical world of tangible things, but also in the conceptions of the people involved.

So, when the robot technology once had been tested in different kinds of environments in the old body shop (first one isolated robot, then a few robots in a line and finally many robots in a line), time had come for a radical socially motivated change in the new body shop in the early 1980's. The engineers were familiar with the robot technique as such and they knew how to handle technical problems. But they did not know how to handle social problems that arose as a result of the initial robotization, namely the degradation of the work tasks in the body shop. They had found that the work force related problems did not disappear, or even change, by installing more robots.

The new body shop, therefore, was planned in order to manage social and technical problems by means of learning and new organization principles. The entire production flow, hence, now was organized in three product shops, each one responsible for its own part of the production. One product shop manufactured the floor, the next manufactured the sides and the third one put the body together to a body frame. In each product shop workers were organized in teams, and the intention from the beginning was to let the workers use their learning capacity so that they can go on developing their skills. All production work as well as routine tasks of preventive maintenance, routine quality control and materials handling should be integrated into the work tasks of the teams. The unions and the company agreed upon this idea in the early 1980's, and there was an initial education for workers and managers respectively. The necessary follow up education, however, was not realized until several years after the inauguration of the new shop. Consequently competence development stagnated and the turn over rates and sick leave figures were not as good as expected and hoped for from the beginning. When the body shop had operated for about ten years, however, these initial problems on the social dimension of the automation process were actively tackled, and the new body shop with its advanced technology and its radical social work organization worked very well in practice, thus solving the technical as well as the social problems of production in a smooth way. Quality, productivity and efficiency improved dramatically. The workers interviewed were satisfied with their extended work content, and their development and learning possibilities. A conclusion is that the process of social change could have been much more efficient, if the power of history, embedded in the minds of people involved, had been met at an earlier stage, more systematically and with recurrent education.

This phase of automation in the body shop can be looked upon as the automation described by Blauner: the more the processing of car bodies resembles on the process of heavy and chemical industry, the less alienated the workers will be. However, this will be true only *if* social measures are taken into consideration in order to change work organization, i.e. to actively change the conceptions of industrial production in people's minds.

5.4.5
Automation in the body shop in terms of the model of change

If we illustrate the automation process in the body shop with a trajectory in the model of change (as principally presented in fig. 1) some conclusions can be drawn, see fig. 2.

First, an U-shaped curve appear at the bottom of the diagram, when only two dimensions are read, namely the technical and the social dimensions. The curve resembles of the U-shaped curve of Blauner (though the U-curve in fig. 2 is turned up-side-down).

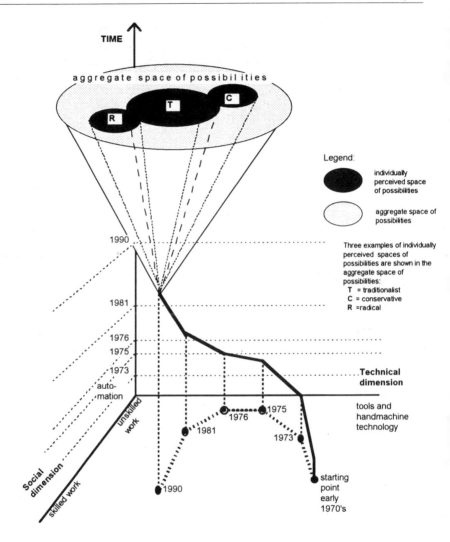

Figure 2   The automation process in a body plant

The trajectory describing the automation process in a body plant from a starting point in the early 1970's and to 1990. From the point in time of 1990 the aggregate space of possibilities' and three 'individually perceived space of possibilities' are depicted. The latter three illustrate three different orientations to the development process: radicals (R), traditionalists (T) and conservatives (C). After 1981 the trajectory moved further on the social dimension towards skilled work.

There is a movement from semi-skilled work with a low degree of automation to a situation where the level of skills is low and the automation increases (the first phase of automation in the body shop example). Next there is a movement towards more automation with no improved skilling (second phase of automation). Finally there is a move towards more skilled work and more automation (the third phase). The concrete content of work tasks, and the kind of skills required, of course differ a lot between the situation when tools and hand-machine production dominated and the situation when integrated work tasks dominate the modern, automated body shop.

Second, the relatively slow pace of change after the period of trying out the robot technology in the Volvo body shop, may be seen as a result of the limited experiences of deliberate social changes in production. The 'individually perceived space of possibilities' of first line and middle managers were restricted as a result of their previous concentration on handling technical changes only, and their efforts to cope only with its negative social effects. They used a defensive strategy towards workers attitudes and were thereby not used to create alternative ways of organizing the work. Now, however, they had to act creatively, to develop an active strategy of how to implement and to behave in a new social organization. It became important to actively avoid negative social effect from the technical equipment and to utilize the potentials of the new technique. This was something new, and it was difficult for them to find their own individual orientation concerning how to handle the new situation. Most of them were for some years, therefore, doing nothing, or very little, they stayed waiting for *someone else* to begin.

People involved in the process of automation described here were used to a situation where technical means solved technical problems. The technical change might, incidentally, affect the organization positively, mainly concerning ergonomic issues. These are important reasons for the slow pace of implementing new social work organization principles.

The new work organization in this case was at the end not looked upon solely as a social means to solve social problems - it was finally concluded that it was a social means to improve the technical performance of the body shop. Therefore, after a lengthy struggle with the prevailing conceptions and habits, the body shop showed successful results.

5.5
Conclusion

To conclude, a general point of this paper is that what initially seemed to be two extremely different conclusions on the effects of automation drawn by Blauner and Braverman respectively, very well may be looked upon as similar conclusions concerning different phases of an automation process seen in a long time perspective. While Braverman concentrated on the mechanization of craft production and identified its negative effects on work content and vocational pride, Blauner focused upon automation in another kind of industry, where the automation process yielded positive effects on the work content.

If we look at Blauner and Braverman from the perspective of automation as an ongoing process two things are obvious. On one hand, Braverman emphasized the early stage of automation, when the process just has begun. The impression of dequalification of workers, and the remaining manual work as residual tasks to what the machine can manage, was overwhelming for him. Blauner, on the other hand, focused upon a later phase of automation. What might be confusing is that conclusions concerning the later phase of the automation process described by Balauner was presented ten years before the conclusions on the earlier phase of automation described by Braverman. It might be a consequence of the different kinds of industry they were engaged in.

In modern industry technical solutions developed in order to avoid unsatisfactory results from older techniques, ceteris paribus, can not be successful if they are not combined with careful social considerations. As the complex problems in modern industry seldom are purely technical by nature, changes must be embedded in the conceptions of people involved. Problems mostly originate from an imbalance, a mismatch, between the new technique and the necessary human and social prerequisites. It is, according to this understanding, impossible to solve complex industrial problems by technical means only without embedding it into its social context. Meeting change like this is one way to extend the 'individually perceived space of possibilities' of individuals, and in an extended time perspective, consequently, the 'aggregate space of possibilities' of the organization will be widened.

Inertia of an automation process emanates from a mismatch between *ideas* of some individuals and what there *really exist* in the physical real world and in the minds of the shop floor actors. A creative engineer might argue that an idea of a technical novelty has great potentials and he describes it as a simple cause and effect situation: given a technical solution the outcome is be predictable. When introduced in a real world workshop, however, the work shop traditions, rules, knowledge, standards of work and skills pattern of workers and managers differ a lot from each other and from that of the engineers who designed the plant. Therein, the inertia of automation is embedded.

If the firm strives to realize the potential profitability of an investment it is important to overcome the inertia of automation processes. Hence, it is not sufficient to implement new techniques without taking the social environment and peoples conceptions into consideration, even if the technique as such in principle seem to be superior to what there already exist. Social obstacles arise when techniques are introduced without thorough preparation and it takes time to overcome them. The time it takes means a great cost for the company. Deliberate social embedding of technical changes, thus, is of fundamental importance to overcome the inertia of automation processes.

5.6
References

Blauner, R:Alienation and freedom. The factory worker and his industry. Chicago university. 1964

Braverman, H:Labor and monopoly capital. New York Monthly Review. 1974

Ellegård, K: Man -Production. Time-pictures of a production system. PhD thesis, Departments of Geography, Gothenburg university, Series B No 72. 1983

Ellegård, K: Utvecklingsprocessen: Ny produkt, ny fabrik, ny arbetsorganisation. Department of Human and economic. Geography, Göteborg university, Choros 1986:5. 1986

6 Automation Strategies at the First-Tier Suppliers in Japan: Process Development and Product Trajectory. Hypothesis on the Supplier-Assembler Relationship[1]

Hisanaga Amikura

6.1 Introduction

The purpose of this paper is to present some hypotheses on assembly automation strategies at "first-tire" auto parts suppliers in Japan. By assembly automation strategy, we mean a coherent set of decisions on building and utilizing capabilities of assembly automation in order to improve manufacturing performance.

It is believed that the Japanese auto parts supplier system has been structurally and behaviourally different from its US and European counterparts, and that this difference contributed to the international competitiveness of the Japanese automakers in 1980s and early 1990s. The structural and behavioural characteristics of Japanese auto parts suppliers are (Fujimoto, 1995):

- Assembler's heavy reliance on suppliers
- The "tiered" structure of subcontracting
- Scale difference in firm size among tiers

---

[1]This paper is an outcome from the research project which is financially supported by The Matsushita International Foundation and The Ministry of Education,Science,Sports and Culture of Japan (Grant-in-Aid for Scientific Research (A) (1) 08303012) . The author is grateful to Prof. Koichi Shimokawa (Hosei University) and Prof. Takahiro Fujimoto (University of Tokyo) for constructive comments.

- Multiple supply relations ("Alps" Structure, Nishiguchi, 1994)
- Long term relationship between assemblers and suppliers
- Suppliers' high technological capabilities (Black Box Parts)
- Competition among first-tier suppliers
- Technology guidance and communication
- Tight link of production systems with assemblers
- Sharing profit from *Kaizen* with assemblers
- Sharing risk of production fluctuation with assemblers.

The "tiered" structure of Japanese subcontracting has been a major source of Japanese automakers' competitive advantage, notably in terms of flexibility and efficiency (Nishiguchi, 1994). To meet the high variety, in unpredictable model mix, Just-In-Time production requirement of their customers, flexibility has been the major challenge suppliers face. However, it is suggested that too much variety hinders assembly automation. Suppliers have to be flexible and efficient simultaneously. How are assembly automatons related to achieving this seeming trade-off between flexibility and efficiency? This is the major research question addressed here.

We will approach this question by comparative case studies, but as of today our study is in process. So the goal of this paper is to discuss the possible alternatives for automation strategy based on the pilot studies, and set the direction for future research.

6.2
Alternatives for Assembly Automation

Fujimoto (1993c) identifies four ideal types of strategies for assembly automation in the automobile industry. The classification is based on differences in the focus of performance that the automation systems are expected to improve.[2]

The first criterion for the classification is the object to be improved. The automation strategies are classified according to whether automation is targeted at product market performance or labor market performance. The second criterion of classification is approaches for performance improvements. One is an element-focused approach, which assumes that superior performance of each element of the automation system, such as each individual automation equipment or work station, should add up to enhancement of total system performance. The other is a system-focused approach, which argues that total system performance is more than a simple sum of element performances, and that basic design or

---

[2] Fujimoto (1993c) insists that the four strategies are not logically exclusive against each other. That is, an actual company can adopt a mix of multiple strategies.

conceptualization of the total system should precede element designs (Iansiti, 1993).

Based on the two-dimensional scheme, Fujimoto(1993c) classifies the basic four strategies for assembly automation: (1) High-tech Automation Strategy, (2) Low-cost Automation Strategy, (3) Human-Fitting Automation Strategy, and (4) Human-Motivating Automation Strategy. (See Figure 1 and Table 1)

Based on this framework, we focus product market performance and technological aspects of automation, and utilize "High-tech/Componential" vs. "Low-cost/Architectural" automation strategies.

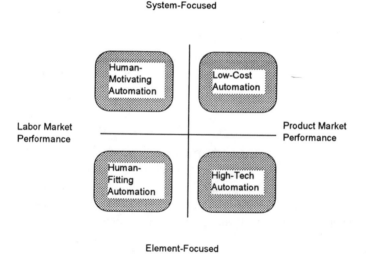

Source: Fujimoto (1993 c)

Figure 6.1 Four types of assembly automation strategy

Table 1 Summary of the Four Types of Assembly Automation Strategies

	High-Tech Automation	Low Cost Automation	Human-Fitting Automation	Human-Motivating Automation
Main Objectives	Competitiveness through advancement of automation technology	Competitiveness through total system improvements	Improvements of physical work conditions at each work station	Elimination of work alienation
Key Measures	Advanced automation equipment for each work station	Total system approach to automation with low cost and limited functions	Automation of "3D" tasks despite increase in manufacturing cost	Automation supporting alternative systems to traditional assembly lines
Strength	Contribution to advancement of automation technology Reputation as high-tech company	Contribution to total performance in cost, quality and flexibility Compatible with continuous improvements of the lean system	Contribution to attractiveness of work place Reputation as "high-touch" company	Contribution to attractiveness of work place Reputation as "high-humanization" company
Weakness	Advanced automation may not contribute to quality and productivity increase Lack of trust between labor and management may be worsened	Assembly work place may not attractive enough to workers	Investment on difficult automation may result in loss of competitiveness through high fixed cost	Utopian pursuit of humanization may result in loss of competitiveness through low productivity
Typical Examples	US and European mass producers of 70s and 80s: GM Hamtramk, VW Hall 54 FIAT Cassino	Japanese mass producers of 70s and 80s Toyota Takaoka Honda Suzuka	Some Japanese producers around 1990 Nissan Kyushu Toyota Tahara	Some European makers under Volvoism influence Volvo Uddevala Mercedes Rastat (?)

Source: Fujimoto (1993 c)

Characteristics of High-tech/Componential automation strategy

The main elements of the high-tech/componental automation strategy are;

• Contribution to advancement of automation technology.

• Technological optimism, or "automation for the sake of automation" mind set.

• Tendency to rely on expensive equipment that may have superfluous functions from total system point of view.

• Tendency to de-emphasize robust equipment design that takes into account future improvements.

• Tendency to buy equipment from outside specialist suppliers rather than making it in-house.

• Top-down equipment design and improvements by specialist production engineers.

• Emphasis on technological "great leap forward" by introducing big automation systems.

The firms pursuing this strategy tended to introduce individual high-tech equipment to the assembly lines without changing the basic concept and design of conventional mass production processes. Such firms seem to believe that advancement of automation technology will almost automatically result in improvements in overall competitiveness of the production systems. We refer to this tendency as "componental."

Characteristics of Low-Cost/Architectural automation strategy

Low cost automation is believed to be an important subsystem of the "lean" production system in the 1970s and 80s. The description of the Japanese makers, for example, case of Nippondenso[3] by Whitney (1992), seems to be generally consistent with the above model.

Fujimoto (1993 c) suggests that this strategy targets total system performance directly, and aims improvement in productivity, quality, delivery and flexibility with simple automation equipment. The main characteristics of the strategy are (Fujimoto 1993c):

• Focus on Overall Competitiveness

Automation is recognized as a means to achieve improvement in competitiveness of the total production system. This approach tries to achieve a given level of total system performance with the simplest, most reliable, and least expensive automation equipment.

• Total System Optimization

Automation is regarded as just one component of the total manufacturing system that includes product design, jigs and fixtures, materials, work design, process flow design, and so on. These factors are simultaneously optimized from total system's point of view, as opposed to designing automation equipment alone without changing the other factors.

• Simple Automation

In order to save investment cost, automated equipment tends to be designed to have "just enough" functions (e.g., flexibility) for the target operations. If semi-automation or power-assist devices are estimated to be more cost-effective, advanced automation technology is deliberately avoided.

• Robust Design

Although the automated equipment may have just enough functions for the current operations, It also adopts robust design in that it is easy to modify or add functions for future changes or improvements.

• In-house Production of Automation

The low cost automation strategy tends to result in higher ratio of in-house design/fabrication of equipment. Standard equipment purchased from outside specialist vendors tends to have superfluous functions for the target operations. Equipment designed and made in-house may also be also easier to improve and maintain for the in-house engineers, supervisors, maintenance workers, and operators.

• Incrementalism

---

[3]The company changed its name from Nippondenso to Desnso Corporation in 1996.

Rather than trying to introduce a big and advanced automation system all at one time, the low cost strategy tends to emphasize incremental approaches of making islands of automation and gradually expanding or connecting them.

• Compatibility with Continuous Improvements

The low cost automation needs to be compatible with the core elements of Toyota-style production system, or the organizational problem solving mechanism. The equipment may be deliberately designed to automatically reveal manufacturing problems, and allow human intervention in response to them.

This strategy has an advantage of continuous improvements, but at the same time a risk of "productivity dilemma (Abernathy, 1978)" also. Nippondenso's Radiator seems to be the case in point. As Whitney (1992) describes, Nippondenso's automation strategy is total system-oriented, and they attack the efficiency-flexibility problem mainly by standardized product design. Traditionally radiators were made of brass, and Nippondenso committed heavily to the brass technology. Calsonic, one of the largest Nissan *Keiretu* suppliers, was considerably faster than Nippondeso in the material conversion into aluminium.

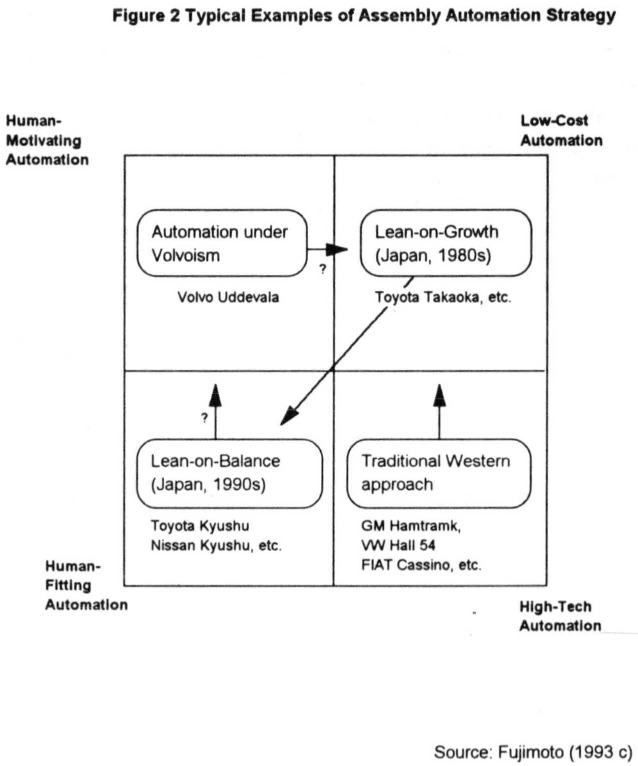

Figure 2 Typical Examples of Assembly Automation Strategy

Source: Fujimoto (1993 c)

Figure 6.2 Typical examples of Assembly Automation Strategy

6.3
Relationship with Assemblers

Fujimoto (1993c)'s taxonomy was for assemblers, not parts suppliers. Although the author believes that the basic logic will be same, there some factors that would be unique to the first-tier suppliers. The typical example would be relationships with assemblers and with second-tier suppliers. They are in the middle of the supply chain, and their strategies for automation may be strongly influenced by the players at both ends of the chain. In the rest of this paper, we focus on the relationship with assemblers.

Congruence between automation strategies

As Fujimoto (1995) points, there are strong linkages between production systems of assemblers and suppliers. Suppliers must meet the high variety, in unpredictable model mix, Just-In-Time production requirement from its customers.

Hypothesis 1
The fit between automation strategies of assemblers and suppliers will increase product-market performance of supplier system.

The automation strategies at assembler influence suppliers' product design itself, desired cost levels of parts, and required volume and delivery timing. To be competitive at product market, assemblers with low-cost automation strategies would not be able to afford to purchase from relatively high priced "high-tech" automation suppliers' parts. Also, Sei (1993) suggests that the higher the degree of automation (e.g., "high-tech" automation at the assemblers), the higher degrees of parts accuracy that will be needed, and the needs for extremely high precision parts will conflict with the movement for parts standardization at the suppliers (e.g. "low-cost" automation at the suppliers).

Hypothesis 1-1
"High-tech" strategy at assembler will fit with that of suppliers, and vice versa.

Hypothesis 1-2
"High-tech" automation strategy at assembler will not fit with suppliers' "low-cost" strategy.

Commitments from assemblers
The pilot study by the author reveals that there are differences in the mode of assembler's commitment to technology development at suppliers. Nissan has "Purchasing, Engineering Support Department" in Supplier Development Group. Purchasing department mainly deals with supplier's technological support. We can call this mode "Outcome oriented" commitment. By contrast, Toyota's commitment to suppliers' technology can be referred to as "Process-oriented" mode, and is mainly done by production technology engineers. As a natural consequence, Toyota's commitment is focused into detailed "one-to-one" cost reduction. Nissan starts to ask suppliers to reduce cost at whole company levels, then negotiates costs of each part.

Hypothesis 2
The fit between automation strategies of suppliers and mode of commitment from assemblers will have positive effects on automation technology development at suppliers.

Hypothesis 2-1
"Process-orientation" of assembler will fit with "Low-cost/architectural" automation strategy at suppliers.

"Process-orientation" of assembler will be consistent with "architectural" automation which would be implemented without changing the basic concept and design of conventional production systems. On the contrary, "High-tech" automation strategy and "Outcome-orientation" commitment will allow suppliers higher degree of freedom to choose and develop original technologies. It would be advantageous for "great leap forward" attitudes.

6.4
Directions for Future Research

This study is at the preliminary stage, and other subjects, such as the relationship with second-tier suppliers, first-tier suppliers' organizational capabilities that make possible concurrent engineering in product development process, capabilities for production technology development, and in-house production facility manufacturing must be studied.

6.5
References

Abernathy, William J. (1978) The Productivity Dilemma: Roadblock to Innovation in the Automobile Industry, Baltimore: Johns Hopkins University Press.

Abernathy, William J., Kim B. Clark and Alan M. Kantrow (1983).   Industrial Renaissance. New York: Basic Books.

Amikura, Hisanaga (1992) "Manufacturing Systems as the Learning Mechanism: Cases from Automobile Industry," Working Paper Series #91M006, Economics Association for Chiba University, 1992.

Asanuma, Banri (1989). "Manufacturer-Supplier Relationship in Japan and the Concept of Relation Specific Skill." Journal of the Japanese and International Economics, 3, March.

Asanuma, Banri and T. Kinutani (1989). "Risk Absorption in Japanese Subcontracting: A Microeconomics Study of the Automobile Industry." Journal of the Japanese and International Economics, 6, March.

Clark, Kim. B. (1996) "Competing through Manufacturing and the New Manufacturing Paradigm: Is Manufacturing Strategy Pass??" Journal of Production and Operations Management 5, no. 1 (Spring).

Clark, Kim B., and Takahiro Fujimoto (1991). Product Development Performance. Boston: Harvard Business School Press.

Cole, Robert E., and Paul S. Adler (1993). Designed for Learning: A Tale of Two Auto Factories." Sloan Management Review, Spring, pp. 85 - 94.

Cusumano, Michael A. (1985). The Japanese Automobile Industry. Cambridge: Harvard University Press.

Cusumano, Michael A. and Akira Takeishi (1991). "Supplier Relations and Management: A Survey of Japanese, Japanese-Transplant, and U.S. Auto Plants." Strategic Management Journal, 12.

DiMaggio, Paul J. and Walter W. Powell (1983) "The Iron Cage Revisited: Institutional Isomorphism and Collective Rationality in Organizational Fields", American Sociological Review, vol. 48 (April), pp. 147-160.

Dosi, Giovanni (1982) "Technological Paradigms and Technological Trajectories", Research Policy, vol. (11), pp. 147-162.

Fujimoto, Takahiro (1992 a). "Why Do Japanese Auto Companies Automate Assembly Operations?" Presented at the Berlin Workshop on Assembly Automation. November, 1992. Research Institute for the Japanese Economy Discussion Paper 92-F-15, University of Tokyo

_____ (1992 b). "What Do You Mean by Automation Ratio?" Presented at the Berlin Workshop on Assembly Automation. November, 1992. Research Institute for the Japanese Economy Discussion Paper 92-F-16, University of Tokyo

_____ (1993 a). "Imakoso Balance-gata Lean Hoshiki Wo Mezase." (Go For the Lean-on-Balance System Now). Ekonomisuto, Mainichi News Paper, March 2, pp. 18 - 25. (English translation is forthcoming from Friedrich-Ebert-Stiftung, Germany.)

_____ (1993 b). "At a Crossroads." Look Japan, September 1993. pp. 14 - 15.

_____ (1993 c). "Strategies for Assembly Automation in the Automobile Industry." Paper presented at The International Conference on Assembly Automation and Future Outlook of Production Systems. Hosei University, Tokyo, Japan. November, 1993.

_____ (1994 a). "The Origin and Evolution of the 'Black Box Parts' Practice in the Japanese Auto Industry" Research Institute for the Japanese Economy Discussion Paper 94-F-1, University of Tokyo.

_____ (1994 b). "Reinterpreting the Resource-Capability View of the Firm: A Case of the Development-Production Systems of the Japanese Auto Makers" Research Institute for the Japanese Economy Discussion Paper 94-F-20, University of Tokyo.

_____ (1994 c). "The Dynamic Aspect of Product Development Capabilities: An International Comparison in the Automobile Industry" Research Institute for the Japanese Economy Discussion Paper 94-F-29, University of Tokyo.

_____ (1995). "Buhin Torihiki to Kigyokan Knakei." (Parts Transaction and Inter-Organizational Relationship) in Uekusa (ed.) Nihon no Sangyo Soshiki (Japanese Industrial Organization), Tokyo: Yuhikaku, 45-72. (in Japanese).

Fujimoto, Takahiro and Takashi Matsuo (1993). "Note on the Findings of the Assembly Automation Study (Second Report) - A Survey of the Japanese Auto Makers" Paper presented at The International Conference on Assembly Automation and Future Outlook of Production Systems. Hosei University, Tokyo, Japan. November, 1993.

Fujimoto, Takahiro and Akira Takeishi (1993). "Jidosha Sangyo no Seisansei (Productivity of the Automobile Industry) ". Soshiki Kagaku (Organizational Science), Vol. 26, No. 4, pp. 36 - 43 (in Japanese).

Fujimoto, Takahiro and Joseph Tidd (1993). "The UK and Japanese Auto Industry: Adoption and Adaptation of Fordism." Paper Presented at the Conference on Entrepreneurial Activities and Enterprise Systems, University of Tokyo Research Institute for the Japanese Economy, Gotenba City, January 1993.

Hayes, Robert H., and Steven C. Wheelwright (1984). Restoring Our Competitive Edge. New York: John Wiley & Sons.

Helper, Susan (1990) "Competitive Supplier Relations in the U.S. and Japanese Auto Industries: An Exit/Voice Approach," Business and Economics History, 19.

Helper, Susan (1991) "How much Has Really Changed between U.S. Automakers and Their Suppliers?" Sloan Management Review, Summer.

Iansiti, Marco (1993). "Real-World R&D: Jumping the Product Generation Gap." Harvard Business Review, May-June.

Itami, Hiroyuki and Tsuyoshi Numagami (1992) "Dynamic Interaction Between Strategy and Technology", Strategic Management Journal, vol. 13 , pp. 119-135.

Jaikumar, Ramchandran (1986). "Postindustrial Manufacturing." Harvard Business Review, November December: pp. 69 - 76.

Jurgens, Ulrich, Knuth Dohse, and Thomas Malsch (1986). "New Production Concepts in West German Car Plants." in Tolliday, Steven, and Jonathan Zeitlin (ed.) The Automobile Industry and Its Workers: Between Fordism and Flexibility: Polity Press: pp. 258 - 281.

Krafcik, John (1988). "Triumph of the Lean Production System." Sloan Management Review, Fall: pp. 41 - 52.

Kumasaka, Hideyuki (1988). "Jidosha no Kumitate Gijutu no Genjo to Shorai" (Current Status and Future of assembly Techniques of Automobiles). Jidosha Gijutu (Automotive Technology), Vol. 42, No. 1: pp. 72 - 78 (in Japanese).

Monden, Yasuhiro (1983). Toyota Production System. Atlanta: Institute of Industrial Engineers.

Nishiguchi, Toshihiro (1994). Strategic Industrial Sourcing: The Japanese Advantage, New York: Oxford University Press.

Sei, Shoichiro (1993). "Contradiction between Standardization and High Quality in Assembly Automation in the Japanese Auto-Industry." Paper presented at The International Conference on Assembly Automation and Future Outlook of Production Systems. Hosei University, Tokyo, Japan. November, 1993.

Shimokawa, Koichi (1992). "Japanese Production System and the Factory Automation." Discussion paper for the Berlin Workshop on Assembly Automation. November 1992.

Teece, David J., Gary Pisano and Amy Shuen (1992). "Dynamic Capabilities and Strategic Management." Revised, June 1992. University of California at Berkeley Working Paper.

Tidd, Joseph (1991). Flexible Manufacturing Technologies and Industrial Competitiveness. London: Pinter Publisher.

von Hippel, Eric (1990) "Task partitioning: An innovation process variable", Research Policy, vol. 19 , pp. 407-418.

Whitney, Daniel E. (1986). "Real Robots Do Need Jigs." Harvard Business Review, My-June: pp. 110 - 115.

_____ (1993). "Nippondenso Co. Ltd.: A Case Study of Strategic Product Design." C.S. Draper Laboratory Working Paper, CADL-P 3225.

Womack, James P., Daniel T. Jones, and Daniel Roos. The Machine That Changed the World. New York: Rawson Associates.

7   Anticipating Problems with Manufacturing during the Product Development Process

Ulrich Jürgens[1]

7.1
The Relevance of Manufacturing Concerns

If asked, almost everybody in industry today would agree that the Taylorist divide between those who plan and structure work and those who execute it has led to adverse effects.   Similarly, there is now a broad consensus that the traditional sequential organization of the product development process, in which manufacturing is the last and least issue in the chain of activities, must be overhauled. The need to anticipate problems with the manufacturability of new products is now widely recognized, as is the need to involve manufacturing representatives in the early stages of the product development process so that they can voice their specific concerns. Only in this way will manufacturing be able to meet the ever increasing challenges of competition over costs, time, and quality.

This assertion is compelling for three reasons. First, opportunities for cost and productivity improvements have long been said to be defined largely in the early stages of the product development process.   Many of the costs incurred by assembly operations result from decisions taken at the product design stage and are almost impervious to measures to improve work organization and the assembly process.   Research in the electromechanical industry and precision engineering has shown that product design accounted directly for only 12% of the total costs of the new products that were studied but was *responsible* for 75%. By contrast, assembly operations accounted for 70% of total costs, but only 13% of the costs can still be influenced at this stage (Gairola, 1985). According to Muschiol (1988), flawed design, often due to insufficient information about

---

[1] Wissenschaftszentrum Berlin für Sozialforschung, Reichpietschufer 50, D-10785 Berlin

manufacturability, explains around one third of the manufacturing costs of new products (Muschiol, 1988, p. 5).

Second, the time needed by manufacturing plants to launch a new product to the point of mass production is becoming ever more critical as manufacturing cycles for models are shortened and the time-to-market for new products is reduced. While many western manufacturers have reduced the number of person-hours per vehicle by introducing lean production principles, they still seem to have major problems achieving the expected level of quality and efficiency during the launch phase of new cars. Start-up problems and resulting shortages of the new, hot models at the dealer's lot prevent the automobile companies from fully exploiting their market potential. Shortening the time-to-market thus has to coincide with shortening the time-to-volume. In other words, manufacturing must achieve a steep product-launch curve.

Recent reports about problems with launches of models by German companies have highlighted the production start-up period as the Achilles' heel of the new product development systems aimed at shortening the time-to-market. The story is similar in the United States, where the Big Three automotive plants still need much longer to change over for new products than their rivals transplanted from Japan.[2]

Third, demand for higher product quality and greater customer responsiveness are increasing. Though quality levels have generally been raised, quality remains a major factor distinguishing new products from competing ones, and many companies are striving to gain or retain a specific competitive edge in this regard. A recent example is the emphasis on clearances and on the fits and finishes of body parts at Volkswagen as part of an effort to improve the optical quality of its cars. This focus creates an enormous challenge to increase the precision and predictability of processing equipment and assembly procedures.

In this article I discuss recent trends and approaches in efforts to take manufacturing concerns into account during the product development process. It draws on research I have carried out in German and U.S. car companies as part of an internationally comparative project on new product and process development networks.[3] In section 2 of this article, I discuss possibilities for

---

[2] The replacement of the '94 Honda Accord in Marysville took three days from the end of the old model to the start of the new model. The '92 Toyota Camry replacement in Georgetown took 18 days; the '95 Ford Contour/Mercury Mystique in Kansas City, 60 days; and the '95 introduction of the Chevy Lumina, 87 days (Treece, 1994, p. 103).

[3] Hereafter referred to as "WZB project." This project, entitled "New Product Development and Production Networks: A Comparison of Countries and Sectors," was coordinated by the Wissenschaftszentrum Berlin für Sozialforschung (WZB) (Ulrich Jürgens, Inge Lippert, and Helmut Drüke) with partner teams in the United States, Japan, and Italy. The project was funded by the Volkswagen Foundation. Research was carried out in the automotive, computer, and machine-tool industries in the United States, Japan, Italy, and Germany (Jürgens, in press).

expanding product engineers' knowledge about manufacturing by means of software in their computer systems or direct work experience. Section 3 deals with the participation of manufacturing workers in cross-functional teams engaged in simultaneous-engineering activities. The rather proactive role of plants in influencing the product and process design is examined in section 4. In section 5 I discuss factors that lie behind the obvious difficulties that many car-makers experience today with their new product launches. The article concludes with a short summary and outlook.

7.2 Providing Manufacturing Knowledge to Product Engineers

Product engineers play the leading part in the evaluation and selection of concepts as well as their elaboration for specific functional solutions and technical specification. Obviously, the most direct way to assure that manufacturing concerns are taken into account is to expand the knowledge and awareness that product engineers have of their role. Two approaches are discussed in this section: (a) the installation of software systems for checking and improving the design for its feasibility in terms of manufacturing and assembly and (b) the use of human-resource development measures that include a period of work in manufacturing in an early stage of the careers of product engineers.

7.2.1
Design Software for Manufacturing and Assembly

Software tools to increase awareness of manufacturing issues during the design stage are reported to be very advantageous in some cases. One example is the design for manufacturing and assembly (DFMA) method (Boothroyd et al., 1993). The Ford Motor Company, for example, has reported billions of dollars in savings as a result of having applied DFMA to the Ford Taurus line of automobiles in the late 1980s (Miller, 1988). The benefits of using DFMA analysis in various industries listed by Knight (1994) are quite impressive (see Table 1).

Table 7.1. Product design-related improvements through design for manufacturing and assembly (from published case studies)

Category	Number of cases	Average reduction (%)
Assembly time	28	58.8
Assembly operations	10	49.5
Separate fasteners	9	74.3
Product cost	9	30.4
Assembly defects	3	28.0

Note. From "The software tool that links design and manufacture" (p. 306), by W. A. Knight, 1994, in P. Zoche (ed.), Herausforderungen für die Informationstechnik (pp. 296-310), Heidelberg: Physika. Reprinted with permission.

Given such potential savings, it seems quite amazing that these systems seemed to receive little attention in the day-to-day practice of most engineers interviewed during the project. Even at an American company that was involved in a precompetitive interindustry project to develop DFMA, very few rank-and-file engineers and draftsmen knew of such a system at all. Not only were the interviewees skeptical that the general manufacturing knowledge encoded in this software could help them with their actual design tasks, they were also unaware of any software that could inform them of the manufacturing requirements of their company. However, DFMA's main application is not to provide engineering solutions to the individual designer anyway. DFMA has proved its usefulness particularly for—

1. training by providing check lists and examples for simplifying the product structure and for suggesting more economical materials and processes,
2. work in multidisciplinary teams by providing a structured problem-solving approach,[4] and
3. reaching a consensus on the best design concept on the basis of the quantified manufacturing cost penalties that can be shown for each design alternative.

Further development of artificial intelligence (AI) could help ensure that individual engineers are provided with specific information about manufacturing conditions. So far, though, progress in the field of AI has been slow and results disappointing. The problem-solving potential of this approach is likely to remain quite limited for some time.

---

[4] Spies (1997) has reported that Volkswagen used DFMA as an integrated way of meeting the various functional requirements in modular product-structuring.

7.2.2
Manufacturing Knowledge Gained through Work Experience

The matter of providing product engineers with manufacturing-related experience as part of their personal development planning receives a great deal of attention at car companies today. The practice is regarded as one of the explanations for the superior manufacturing performance of Japanese companies. It finds particularly strong support at the plant level, with some interviewees emphasizing it as a sine qua non for improving the manufacturability of design solutions. Many product engineers and draftsmen used to have at least some exposure to manufacturing environments early in their professional career. However, recruitment patterns have shifted in favor of university graduates who have had no manufacturing experience and who are primarily interested in working with computers.

Despite the enthusiasm about personnel development systems that promote job rotation and manufacturing experience, actual practice is quite different. None of the Western companies in the research sample had established such a system for new entrants in product engineering. If recruitment took place at all, the perceived need for these engineers was too great in the engineering departments to have the newcomers "waste" their time on manufacturing tasks. The only systematic approach to include manufacturing experience (as well as experience in other functional areas) was observed in special trainee programs for high-potential individuals, mostly specially selected university graduates. Such programs included intervals of actual manufacturing work. In general, the rank-and-file engineer remains outside border-crossing rotation and training programs. However, various pragmatic and ad-hoc initiatives do exist to reduce the mental distance between product engineers and manufacturing. At one German car manufacturer, product engineers who have already specialized in a certain component field are encouraged to spend some days familiarizing themselves with the production area while these parts are being manufactured and installed. At present (1998) around 5% of the engineers at the company's product development center have had such experience. At another manufacturer, some departments in product development had organized "production days" during which the staff worked on the production lines.

Even such short exposure to actual working conditions were described as quite fruitful. Product engineers who had had that experience particularly emphasized that they could now more easily relate to manufacturing representatives in cross-functional team activities, could understand and anticipate their concerns, and tended to visit and contact the manufacturing site more often than product engineers who had not familiarized themselves with that area. This assessment was corroborated by interviewees from manufacturing as well. Department managers in product development, however, were torn between the long-term benefits of human resource development and the need to cope with the current workload. Time-to-market pressure, downsizing of staff, and an increased number of development projects clearly relegated considerations of personnel development to a low priority.

7.3
Representing Manufacturing in Cross-Functional Teams

Basically, the literature on product development differentiates between two ways of organizing the product-to-market process: the organization is either functional or product-oriented, with the implication being that the rules of the game are either set by the functional heads or by some kind of product management. Research on this project suggests that a purely functional organization of the product-to-market process no longer seems to exist in practice. The sample companies were all hybrids of functional orientation *and* product orientation. A matrix scheme, which is usually referred to in such cases, does not greatly help capture the important differences between these hybrid forms. All companies in the sample had introduced cross-functional teams as part of their reformed product development systems and as the organizational prerequisite of simultaneous (concurrent) engineering.[5]

The question in the following pages is how manufacturing is represented in the activities of the cross-functional teams. I first examine cases in which functional orientation dominated product orientation. All of these companies had introduced cross-functional teams by the mid-1990s. In one German company a structure had been created with a general cross-functional steering team at the executive level, approximately 20 teams responsible at a meso-level for different areas of the car, and subordinate operational teams responsible for subsystems within each module. Altogether there were around 60 such "chunk" teams for one new product project.

Manufacturing was represented on all of the teams at the meso-level and in most of the chunk teams. Normally, the manufacturing representatives were process engineers from production planning or production supervisors temporarily assigned to this task. Representatives from the body shop would attend the team activities related to sheet metal design; representatives from assembly would deal with exterior and interior trim issues, and so on. The teams primarily had the task of *coordinating* activities carried out within the functional areas. They were institutions for information exchange. The team leaders were generally senior product engineers from the design department, although this arrangement was not formally stipulated. The team members were delegates

---

[5] The terms are used synonymously here. In the North American context the term *concurrent engineering* is preferred, in Europe the term *simultaneous engineering* is more established.

from their respective functions. The decisions made by the team could be overruled by the management of the functional area.

Though superior to the previous approach of sequential development in functional isolation, the delegative approach seems to be only a small step toward integrating design and manufacturing. The cross-functional teams are regarded as a manifestation of simultaneous engineering, but within the company most of the teams remained far from this ideal in their actual work. Often the team leader saw his (or her) primary task to be that of implementing the design concept and assuring its acceptance by the other team members. The members of other functional areas often had trouble attending the various team meetings, providing feedback to the relevant groups within the affected functional area, and listening to their colleagues. For example, the delegate for the assembly operations in one of the sample companies represented several component areas and many chunk teams. His role as representative was a full-time job, yet he would often send other persons in his place because of his tight schedule. Of course, he would inform the assembly areas, seek opinions, and inform others of decisions, but feedback and consultation meetings in production usually involved the manager and supervisory staff only. Consequently, most of the supervisors and workers received little information about ongoing development of new products and were confronted with its result only at a very late stage. It is thus no wonder that the demands and the pressures of introducing a new product were regarded by the work force as an interruption of their regular work rather than as a chance to enrich their work and create opportunities to involve them in optimizing the process.

The link between the involvement of experts and the involvement of the rank and file in the new product process was a crucial issue. Only when the project reached the point at which a pilot car was built were supervisors and a selected group of production workers confronted with the new car model. Most of the work force did not have a clear-cut idea of the new car and the related changes until the first test cars went down the production lines in their plants.

The other German car manufacturer in the sample had developed a more structured approach to linking the involvement of the manufacturing area's process planners and their rank and file colleagues with the product development process within the plant. It was related to the introduction of teamwork in the production areas of this company. At the beginning of series development, a project leader would be nominated to coordinate plant involvement in the project. In coordination with the function managers, the project leader would nominate the delegates for the various cross-functional groups in charge of the development of the car's subsystems at the development center. These delegates were normally process planners or quality experts. (The number of cross-functional teams in this company was usually much smaller than in the previously described company because larger task chunks were assigned.) At the same time a "multiplier" for each production section would be nominated. This person was in charge of voicing shop-floor concerns at the development center, participating in the construction of the prototype, and feeding information back and forth to his or her production area in the plant. When the pilot stage was reached, the number of multipliers increased to one per production team. These persons would often be the team leaders (in this company, a non supervisory position whose incumbent was elected by the team members) or experienced workers who could master all tasks within the team. The number of multipliers who spent most of the week at the technical center during this time would rise to 10% of the production work force at this stage.

The multipliers would collaborate closely with the process planners sent from the plant into the cross-functional teams, which were in charge of engineering the specific components at the technical center. In this way process experts and shop-floor workers were able to represent plant interests forcefully while bringing their knowledge and experience to bear. A considerable number of regular shop-floor workers became involved in the program at an early stage. Additional training measures and a series of public events instituted when the program was handed over to the production plant rounded out the picture. All in all, the procedure that was developed together with a major automobile program in the company from 1992 to 1995 was regarded as a success and has become standard procedure at the plant.

The alternative to organizing projects by function is to assign them to "heavyweight" project managers (Clark & Fujimoto, 1991) or a separate management unit responsible for an entire product line. The German companies in the research sample favored a functional orientation, whereas the American sample companies had segmented development by product line. This project-centered approach tends to set higher standards of cross-functional teams, providing focus (the team members usually have no other responsibilities), empowerment (project management has authority in personnel and budget matters), and simultaneous engineering (parallel work by and intense interaction among produce and process engineers within and across company boundaries). The feasibility of this mode of organization depends largely on co-location, an arrangement whereby all major partners involved in the product-to-market process are located at least together at least part of the time with their platform team. In this WZB project, spatial proximity was found to be the most important factor in improvement of communication and cooperation in the process chains (see Figure 1).

In one of the U.S. car companies the members of the manufacturing staff assigned to the platform were located just on the other side of the floor, so the involvement of manufacturing did not require much time and organizational effort. As one product engineer explained: "If you ever have any issue, you just have to go looking for them. If you can't find them, leave a note on their chair." The result was that the manufacturing representatives were consulted more often by the product engineers and were regarded less as watchdogs than was the case under other conditions.

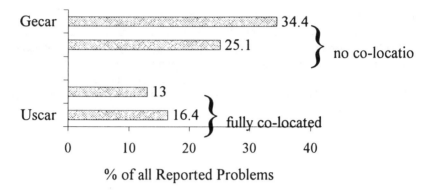

% of all Reported Problems

Fig. 7.1. Spatial distance as a cause of problems of communication and cooperation with other functions in the product-to-market process chain at car manufacturers with and without co-location (in percentages).*

*Departmental managers and the rank and file in each of seven in-house functional areas were asked to note problems with their process partners of the other functional areas and with three types of suppliers. They were also asked to give the main reason for each of the problematic relationships. Spatial distance was one of eight possible problem-causing factors. The four companies were major car manufacturers. The countries represented were Germany, Italy, and the United States.

Source: Project on New Product and Process Development Networks, 1993-1996, Wissenschaftszentrum Berlin für Sozialforschung.

As with function-dominated project organization, the representatives of manufacturing are process engineers delegated by the manufacturing function to the platforms. The spatial proximity between product engineering and production planners or process engineering does not eliminate the potential for interface problems between the co-located engineering groups and the production plants. Thus, the danger is that a gap will open between the production planners and the production-operations level. Nor does co-location guarantee the possibility of direct interaction and face-to-face consultation at any time. With their widened responsibilities, product engineers and process planners, who are also in charge of coordinating the work of engineering service companies and suppliers, will be out of their offices much of the time. Their ability to allow for immediate communication and cooperation thereby remains a problem even if co-location is practiced.

The three cases described in this section differ particularly in the way the shop-floor work force is involved in the process of new product development. In the first example, involvement was minimal and did not begin until a prototype was built.   In the second example, a structured approach to linking the involvement of process planners and that of the plant's shop-floor workers was created to allow for more forceful representation of the production plant's interests at the product development center.   In the third case, shop-floor participation in the early stages of developing a new product had an almost campaign-like character in the plant. In the second and third examples, the companies had created formal and informal organizational structures that smoothed the flow of information between the production plant and the product development center and between shop-floor workers and process experts within the production plant.   The overlap and redundancy of information certainly was an additional cost factor, but as the literature on knowledge theory shows, information overlap and redundancy is precisely the way to foster organizational learning and build the "knowledge-creating company" (Nonaka, 1990; Nonaka, Takeuchi, 1995).

The passage of the new-car project from the prototype stage through the pilot stage and into production is a particularly instructive example of how organizational learning can be promoted or inhibited.   In large car companies the prototype stage and prototype-building, which have to do with development activities and testing, are traditionally separate in both time and space from the pilot-building stage, which deals with manufacturing activities and process optimization.   In some companies these two stages, even their respective work forces, still are completely separate, with little job mobility between the two. Natural learning paths become organizationally divided.   Prototype-builders, who are the first to put the new car together, have no knowledge of the actual conditions in the production plants, and the manufacturing staff receives no information about the experience gained as the prototype is built.   Whereas it is true that the main function of prototypes is to test and confirm the work of the product designers, the function of the pilot-building phase should be to check for manufacturing concerns, optimize the work process, and train workers.   In practice, however, much of the pilot-building activities are still related to prototyping.   The conditions of mass production can hardly be anticipated, and only a few workers from manufacturing plants participate in the building of the pilot cars.   As will be shown in the next section, an increase in the share of the work force involved in pilot-building and an intensification of their involvement can greatly facilitate the introduction of a new model and, particularly, for the involvement and motivation of the manufacturing work force during and after launch.

7.4
Mobilizing the Production Plant for Launching the New Product

There is not only an increased need to take manufacturing concerns into account early in the process of developing new products, there is also an increased preparedness and will of the manufacturing work force to take on a more active role. Two reasons for the proactive stance of production plants are discussed in this section. I then describe a representative example in which the management and workers of an assembly plant took initiative and fully mobilized their knowledge in order to ensure that the new product became a success.

The first reason for a more proactive role of the production work force lies in what one might call the "humanization dilemma." This observation pertains especially to assembly workers. On the one hand, these workers are much more carefully selected and much more qualified than in the past and thus have much higher expectations of their job content and careers. On the other hand, attempts have largely failed to humanize work through job enlargement and enrichment. In fact, there seems to be a renaissance of the classical short-cycle assembly line, and many of the advanced new factory layouts have been scrapped (see Jürgens, 1997). As a way of fully using the potential of members of the work force and meeting their expectations, the activities entailed in developing new cars seem to be well-suited to enriching jobs and involving workers. The argument is not only one of motivating people but also of enhancing the efficiency with which human resources are used. As the cycles of models shorten and as the concomitant variants and options increase, so does the amount of work related to engineering changes and new product introductions. An educated guess is that around 20% of the total time spent by the average assembly operator in today's assembly plants over a five-year period is devoted to dealing with such activities (including engineering changes, model updates, and the introduction of new models). This work is irregular and neither standardized nor optimized by work planners. It is communication-intensive work that takes place in close cooperation with product and process engineers and internal and external suppliers. It might even include contact with customers. Thus these activities mean a respite from short-cycle repetitive standardized work and could serve as a welcome source of tasks for enriching assembly work.[6]

---

[6] This approach is explicitly formulated by a Northern European car manufacturer. In policy guidelines regarding the company's "future plants," the corporate management has stated its willingness to "stimulate teamwork and cooperation" and "develop teamwork to include production, drafting, and design work" (1995 company document for future work organization and production systems). As a first step toward implementing this vision, the company introduced personnel development systems and new job profiles aimed at increasing integration and networking between product development and manufacturing. When I visited the plant in 1996, some of the

The second reason for the work force's more active role in the process of developing new products is related to the organizational changes moving toward simultaneous concurrent engineering. These new approaches are designed primarily to bring product engineers and process planners into close interaction underscored by their co-location. For production plants this arrangement entails the danger that their own already minor role in the product development process will diminish even further. After all, "their" process planners become the partners, possibly the allies, of the product engineers and may side against the production plants when interests in product development collide. This perceived threat comes at a time when management and workers at production plants feel increasing pressure from globalization and in-house competition. The performance of plants in this product program some years hence may determine whether they will win the order for the next program in the bidding against other plants. Thus there is the feeling that all the resources of the plant will have to be mobilized to make this program a success.

The following episode illustrates a case in which the work force of a local plant took the initiative (for a more detailed account see Haddad, 1998). The assembly plant where the new model was to be launched was located several hundred kilometers from the simultaneous engineering center where the model was being developed. Prototype-building and pilot-building were also co-located at this site—on two sides of the same development center. As usual, plant involvement in the concept phase included mainly process engineers from the production planning and quality-control departments. But groups of workers participated in design reviews as early as the clay model stage. During the prototype stage, the number of workers sent to the concurrent engineering center increased, peaking when the pilot stage began. These "guests" who came in groups with their supervisors and with process engineers for at least one week, worked on the pilot-building process, visited suppliers of parts and process equipment, and developed their suggestions for improvements. Altogether, about half the work force of the assembly plant, some 1,500 people, had worked at the company's distant development center before the preproduction stage began in the plant itself. In order to check devices and develop alternative solutions that tapped the knowledge and experience of the production workers as pilot-building proceeded at the local plant, space in the plant's assembly area had been cleared and equipment installed to emulate some of the processes at the development center. This area for in-house pilot-building was regarded by plant management as a first

---

assembly areas had already changed the organization of their teams. Within the team the position of "product and process support specialist" was created. This specialist received special training and served as the team's liaison with the process planners and product engineers. Two "specialists" per team were assigned, and these assignments rotated among the team members in order to give everybody the opportunity to develop their abilities.

step toward shifting more of the pilot-building function into the assembly plant when the next model reaches the phase of prototype and pilot-building.

When I visited the plant in the course of the project, the stage of mass production had already begun.[7] The work force's strong sense of ownership, which was induced by this intense involvement in the new product, was still palpable, with pride extending to the highest level of local union leadership. Special conditions at the plant had obviously made it possible to involve a relatively large number of workers, an opportunity that would have been much more difficult to come by in busier times. But the principle of early and broad participation of the assembly plant's regular workers, who served the suppliers and the prototype- and pilot-building operations as experts, has proved to have great merit. The interviewees' overall reaction to their involvement ranged from positive recognition to great enthusiasm, as expressed by the following statements from our interviews. The first was spoken by a product engineer, who confirmed that the positive experience was not just on the manufacturing side: "I cannot quantify the benefits. It's a major improvement. We were able to bring the assembly line up faster and with fewer people than in any program I have ever been on." This early involvement provided a valuable network for launch preparation. A product engineer commented:

> The biggest difference compared to previous launches was that the people in the plant had been members of the group, the team, for the prior two years. They knew about everything, and there really were not any surprises. . . . . They knew what was going on. We knew them, we knew not just the launch people, we knew many more people in the plant, just based on all the contacts over the years. And they just made it so much better. And you also knew more about what was going on in the plant than you would have had in the prior launches. It was not so secretive. Before, we had the engineering people here and the manufacturing people there. And you didn't really know the manufacturing issues. They would not share them with you.

The clash between the worlds of product development and production, which had been separate for such a long time, must have been considerable. Now, however, both groups seemed to enjoy working together and emphasized the advantages of doing so. No wonder there was so much pride about this process at the plant level. Said one process engineer from a perspective at the plant level:

> I thought I really was in tune with the line, but those guys, they do it 400 times, and they've got their own rhythm and they have got things they come up with. . . . There were really a lot of surprised engineers at the center when these people showed up. . . . There were FMEA meetings—those were really eye-opening for a lot of people. A lot of good suggestions came out of that. You

---

7 This part of the research was carried out with Carol Haddard (University of Eastern Michigan).

would be surprised to see how many questions the engineers from central staff asked. The product people were willing to accept constructive criticism and really did a good job once they accepted it. The hourly people also learned the engineering view of it. We would show them what it takes to make the changes they asked for. We took them to the vendors. And they found out why some things were done, that some things are safety items, whereas down the line you think, why do we need to do this?

7.5
Reasons for Troublesome Launches

The factors hitherto discussed—designs that do not lend themselves well to the manufacturing process, blocked natural learning paths, untapped knowledge and motivational resources, non-empowered teamwork, and so forth—partly explain the problems that ultimately surface in production plants when it comes time to launch the new product. In fact, it seems that the degree and intensity of problems at launch time have been increasingly since the early 1990s. Three factors bearing on these complications are briefly discussed below.

1. *Loss of manufacturing competence.* The attempt to develop everything new at the same time (a new car model, a new manufacturing technology, new production control and logistic systems, and new organizational structures) is an effective recipe for a crash and can often be heard as an explanation for past failures in Western car companies (Bungard & Hofmann, 1995, p. 22). Few resolutions have been forgotten more often than that of never making such a mistake again. European manufacturers in particular have continued to introduce completely new process concepts together with new product programs. This practice has been related to their new manufacturing strategies, which aim at—
   – the modularization of the product as part of new strategies of systems suppliers and platforms,
   – the use of new materials such as aluminum for body parts, and
   – the adoption of Japanese-style continuous-flow production principles.

In this way projects for developing new products became part of change dynamics, the result of a particularly intense effort to manage change. Confusion was compounded by the loss of old hands, experienced managers, and staff in the wake of sweeping early retirement programs and by the loss of expertise in process technology because of downsizing and outsourcing. Current product development projects consequently face a dilemma: Company strategy demands an approach that integrates product and production concepts, but the development of such concepts has been outsourced to a large extent. This outcome has encroached upon another leading principle underlying state-of-the-art management of product development: *frontloading* (Fujimoto, 1998). Because of downsizing and outsourcing, more and more companies have turned to external firms even for concept development of the new manufacturing processes that are intended to be introduced with the new products. The process of writing a tender, receiving bids, and selecting the desired firm consumes precious time and in many cases frustrates all good intentions behind integrated concept development.

2. *The "downstreaming" of unfinished tasks to the preproduction and launch stage.* With shortening time schedules and rising pressure for time and cost reduction, the risk is growing that problems with unfinished tasks are accumulating in the downstream phases of the development process. This danger is particularly great for tool-and-die manufacturing, specifically the passage from prototype-building to pilot-building and then to production launch. Consequently, manufacturing has little chance to establish reliable, robust processes optimized to produce according to the specifications developed by the product designers. This tendency can be observed in the pilot stage, when all parts should, in principle, be produced by mass-production tools and dies, a point that marks the end of the product-engineering work and the beginning of production preparation. The reality is starkly different. Many parts, including a large number still at the preproduction stage in the plant, come from preliminary processes in which many manual corrections and adjustments are made to both the components being manufactured and the tools and dies producing them. It is impossible to optimize production processes under these conditions. Worse yet, the process equipment for mass production at the plant has to be broken in and adjusted so that it can deal with parts that come out of preliminary processes. Such adjustment frequently requires the delicate removal of metal here or filling in of metal there, intricate operations often executed by the local maintenance staff. When the number of parts coming from more fully optimized production processes later increase, the equipment has to be readjusted. In the meantime, other parts have matured and begin triggering new changes. Even without design changes from product engineering, manufacturing activities seem to go through endless ripples of change in this way. The ad hoc and uncontrolled character of changes leads to the progressive loss of the original database, which cannot be reinstalled without considerable measuring effort. All attempts to produce parts that strictly conform to design specifications are frustrated. The tendency to lose the database because of incomplete homework in previous processes accounts for many delays and for the intensity of the problems that have beset many recent product launches.

3. *The lack of a controlled approach to launching new products.*[8] The preproduction and launch period is an intense and difficult time for manufacturing. Problems abound, change ripples back and forth, sudden events require immediate response. These conditions require fire-fighting on many fronts. Indeed, many plants have developed great fire-fighting capability. However, they often lack the ability to improve processes in a gradual and controlled manner, an approach that requires the whole shop floor to check parts continuously, monitor tolerances closely, and record improvement and deviation. Whereas effective fire-fighting is possible through reliance mostly on experts, a gradual and controlled way of dealing with problems requires the broad involvement of the shop floor. It also necessitates metrics, benchmarks, procedures based on experience with previous launches, and a work force with high problem-solving potential (e.g., statistical process control capability). Such capabilities were not asked for in the era of Taylorism, however, and the heritage of that era particularly hurts during the launch period.

7.6
Conclusions

My conclusions can be summarized in three points:

1. As cycles for models become shorter, and as the time-to-market of new products becomes briefer, the complications of manufacturing mount. The difficulties due to poor manufacturing design and uncompleted upstream tasks accumulate in the phase of preproduction and launch.
2. There is no readily apparent quick fix for this problem. A solution can come only from a mix of measures that includes training and personnel development for product engineers, engineering tools for computer-aided design to improve the manufacturability and assembly of products, organizational measures that bring manufacturing staff into the early processes of developing new products, and a structured approach that provides for a controlled product launch and enables manufacturing workers to set up their equipment strictly according to specifications.

---

[8] The superior product launch capability of Japanese production plants was impressively shown by Clark and Fujimoto (1991, pp. 188–204). This superior capability can be seen as both the prerequisite for and the consequence of the fact that the Japanese companies tend to minimize the phase of pilot-building in a separate pilot plant, even dispensing with that phase altogether and building the pilot cars on the existing volume line right from the outset. By contrast, European manufacturers prefer to have separate pilot plants and start building the new cars on existing volume lines relatively late (p. 190).

3. The new product development systems that aim at integrated and parallel cross-functional work offer real opportunities to take manufacturing concerns into account during the development process. The adoption of simultaneous engineering methods should, however, aim at broad involvement of the rank and file at the production plant. This approach would benefit from the specific expertise and knowledge of these workers while offering new opportunities for job enrichment in production.

7.7
References

Boothroyd, G., Dewhurst, P., Knight, W. A. (1993): Product Design for Manufacture and Assembly, Marcel Decker, New York..

Bungard, W., Hofmann, K. (1995): Innovationsmanagement in der Automobilindustrie (Innovation Management in the Automotive Industry), Beltz Psychologie-Verlagsunion, Weinheim.

Clark, K. B., Fujimoto, T. (1991): Product Development Performance. Strategy, Organization, and Management in the World Auto Industry, Harvard Business School Press, Boston.

Fujimoto, T. (in press): Shortening Lead Time through Early Problem-Solving—A New Round of Capability-Building Competition in the Auto Industry, in: Jürgens, U. (ed.), New Product and Process Development Networks, Springer, Berlin.

Gairola, A. (1985): Montage automatisieren durch montagegerechtes Konstruieren (Automating Assembly through Design for Assembly), in: VDI-Zeitung, vol. 127, no. 11, pp. 403–408.

Haddad, C. (in press): Involving Manufacturing Employees in the Early Stages of Product Development: A Case Study from the U.S. Automobile Industry, in: Jürgens, U. (ed.), New Product and Process Development Networks, Springer, Berlin.

Jürgens, U. (in press): New Product Development and Production Networks, Springer, Berlin.

Jürgens, U. (1997): Rolling Back Cycle Times: The Renaissance of the Classic Assembly Line in Final Assembly, in: Shimokawa, K., Jürgens, U., Fujimoto, T. (eds.), Transforming Automobile Assembly. Experience in Automation and Work Organization, Springer, Berlin, pp. 255–273.

Knight, W. A. (1994): The Software Tool that Links Design and Manufacture, in: Zoche, P. (ed.), Herausforderungen für die Informationstechnik, Physika, Heidelberg, pp. 296–310.

Miller, F. W. (1988): Design for Assembly—Ford's Better Idea to Improve Products, in: Manufacturing Systems, March 1988, pp. 22–24.

Muschiol, M. (1988): Rechnerunterstützte Informationsbereitstellung für den Konstruktionsprozess am Beispiel montageorientierter Gestaltungsrichtlinien (Computerized Supplier of Information for Design of Assembly Systems), Hansa-Verlag, Munich.

Nonaka, I. (1990): Redundant, Overlapping Organization: A Japanese Approach to Managing the Innovation Process, in: California Management Review, spring 1990, pp. 27–38.

Nonaka, I., Takeuchi, H. (1995): The Knowledge-Creating Company, Oxford University Press, Oxford, England.

Spies, J. (1997): Montagegerechte Produktgestaltung am Beispiel des komplexen Großserienproduktes Automobil (Product Design for Assembly as Illustrated by the Complex Mass-Produced Automobile), Doctoral dissertation, Eidgenössische Technische Hochschule Zürich.

Treece, J. B. (1994): Motown's Struggle to Shift on the Fly, in: Business Week, July 11, pp. 102–103.

8 Diffusion patterns of lean practices: lessons from the European auto industry

Arnaldo Camuffo and Anna Comacchio[1]

8.1 Introduction

Building on the original work of the authors and other international researches (MacDuffie and Pil, 1996; Kochan and Lansbury, 1996, Shimokaua, Jurgens and Fujimoto 1997), this paper examines some relevant aspects of automation and HRM in the auto industry after the "lean" revolution (Womack, Jones and Roos, 1989).

The paper analyzes the diffusion process of "lean" technology, HRM-IR and work organization practices and argues that the search of competitive performance in terms of efficiency, quality or flexibility-responsiveness to market changes is a systemic and dynamic pattern. From this standpoint the discussion focuses on two issues: diffusion process and human resource management role.

Firstly, it is undeniable that the search of competitiveness is forcing automakers to adopt automation and organization strategies tending towards "lean" solutions. But these solutions are interpreted and applied by firms in rather different ways. The drivers of these diverging experiences are the internal adoption process and the external institutional and economical constraints. The firms simply do not imitate but rather they enact a specific, contingent creative combination of firm policies and new techniques which result in a variety of technological and organizational models. Moreover, as far as the adoption

[1] The author would gratefully aknowledge CNR (Italian National Research Council), Murst (Ministry of University and Scientific and Technological Research) and IMVP for their research funding. Although the chapter is the the result of a common research work, Camuffo has written the following psaragraphs: 8.2, 8.4, 8.6, 8.7, 8.9 and Comacchio the following ones: 8.1, 8.3, 8.5, 8.8

process is concerned, both the institutional and competitive context are important.

Secondly firms' "lean automation strategy" is only one competitive leverage. HRM strategy counts both in terms of "social matching" between automation and organizational models and in terms of "lean employment strategy".

As a case study, the paper uses evidence from the European auto industry to show how the "lean" concept tends to evolve as it is implemented in different contexts (and at different times). Strategies of adjustment to "lean employment" in the European car industry are identified and explained as contingent on external (institutional) and internal factors.

8.2
Lean management system: some defining issues

It is undeniable that Japanese manufacturing techniques and organizational practices represent a substantial evolution of the mass production paradigm. Adoption and implementation of lean production principles led to relevant improvements in the industrial performance of auto makers.

By now there is a considerable body of literature analyzing how the Japanese management system was founded and how Japanese firms developed specific features partly in contrast (Womack, Jones and Roos, 1989) and partly coherently (Coriat, 1991) with Fordist principles. More recently, a number of studies tried to understand dynamics of the diffusion of the lean manufacturing concepts, although prevalently focusing on the hybridization of this model through the transplants in United States (Kenney e Florida, 1991) and Europe (Garrahan and Stewart, 1992). These contributions shed more light on both the characteristics of this management system and its reproduction in different institutional and cultural contexts. At the same time, implications of "lean management" are becoming wider (Womack and Jones, 1996) and tend to cover the whole auto supply chain.

Some argued that the debate on the nature and diffusion of lean production seems to be outdated. Recent research from Japanese scholars (Fujimoto, 1994) show that auto manufacturing systems are rapidly evolving even in Japan, and that the "lean production model", as designed in the original book, is probably no longer the best theoretical benchmark so to understand the dynamics taking place. This is the reason why scholars move "beyond lean production" (Kenney and Florida, 1993) and emphasize the determinants of intra-firm and inter-firm evolutionary dynamics in order to map the emerging patterns.

However, it must be underlined that implications of "lean" revolution are far from being completely worked out. If the extent of changes implied by "lean" paradigms are comparable to "Scientific Management" or "Fordism," it is easy to discuss that there is still much to study about applications and implications of "lean manufacturing". This task will take much longer than six years.

From this standpoint, organizational, industrial relations and HRM issues are even more peculiar. In fact, they were not fully developed in the original "lean" book and thus the margins to increase understanding in these areas are even greater.

Moreover, IR-HRM variables are important as a field of study in auto manufacturing because "lean practices" in this area remain a rather fuzzy concept, in the sense that they cannot be unequivocally defined. For example, flat organizational structures have often been pointed out as main features of lean manufacturing, while recent field research (Ittner and MacDuffie, 1994) show that the number of hierarchical layers in Japanese auto plants is, on average, greater than in less "lean" (e.g. European) contexts (see Table 1).

Table 8.1   Plant differences in the number of organizational levels

Region/Plant	Year	Number Of Organizational Levels (Sample Means; Actual Data)
JPN/JPN	1993	7.9
JPN/NA	1993	6.5
US/NA	1993	6.2
EUROPE	1993	7.1
SEAT Martorell	1995	6.0
FORD Valencia	1996	5.0
HONDA Swindon	1996	5.0
RENAULT Flins	1996	6.0
FIAT Melfi	1996	6.0

Number of organizational levels from production workers to plant manager

Source; Ittner and MacDuffie, 1994; EIU, 1995, 1996; plant interviews

Furthermore, there is no univocal matching between the "lean" concept and specific HRM or industrial relations policies. In fact, while some authors lay the stress on the "mean" nature of lean production and its connection with unilateral or concession bargaining models of IR (Robertson and Others, 1992; Cerruti, 1995), others underline the possibility to join elements of the LP model and co-operative, partner-based union-management strategy (Kochan, Bennett and Rubinstein, 1994).

This ambiguity comes from three factors. First of all, the nature of these issues (psychological, social, and institutional variables largely differ across individuals, regions, countries and so on) does not allow an immediate translation of concepts and principles in approaches, systems, techniques, as can happen when talking about logistics or operations management. This implies that while electronic kan-ban or TPM are sufficiently well defined and codified, a team, a suggestion, a competency, a job grade or rank, an incentive (contingent compensation) or absenteeism cannot be easily defined and measured without immediately verifying country-specific, region-specific, firm-specific and even plant-specific factors.

Secondly, there is a measurement issue. While metrics for manufacturing efficiency (for instance in terms of cost, quality and delivery) are easy and solid (in part due to objectivity and in part to longer tradition in performance measurements), metrics in terms of HRM and IR are comparatively less developed and somewhat vague.

Thirdly, this is not only a "technical" issue, but also a policy issue, as choosing measures reflects choices in terms of value. For example, if the "voice of society" (for instance in terms of maintaining employment levels or workers welfare) must be taken into account so to evaluate overall performance of the "lean" model, then a wide range of IR-HRM metrics ought to be used.

On the whole, these factors imply that inter-plant or inter-firm comparisons as well as to evaluate the distance of real situations from supposedly "lean" standards it is easier and more practical for manufacturing efficiency. This is extremely difficult and sometimes concealing for IR-HRM systems.

8.3
Convergence and divergence in automation practices and the human resources role

As sustained by the seminal work of Nelson and Winter (1977), one fundamental dimension of the natural technical trajectories is the increasing mechanization of operations. Automation, considered as replacement of workforce with equipment, robots, computers, etc., is a long term pattern. In the automobile sector the continuous substitution of human intensive processes with capital intensive activities has been a secular trend. Considering the last decades, automation seems to be an on going progress, and recent research shows that, even in the last few years, the overall automation level has increased moderately (EIU 1996, EIU 1997a). This long term progress of automation, defined as a technological trajectory, is the "movement of multi-dimensional trade-off among the technological variables" which is considered relevant (Dosi 1982). This concept reveals the intrinsic nature of technological innovation as a learning process within a multi-dimensional space mainly defined by physics and economics (Whitney 1997).

Considering the economic space, the automation trajectory is the combination of problem solving activities (analysis, decisions, investments) driven by four main variables (Fujimoto 1997): costs (economy of scale and labor cost reduction); product quality (especially concerning body welding and painting operations); flexibility (static adaptation of equipment to model variation and dynamic capacity to assemble different models with the same line or reduce lead-time changeover and lower investments for new model introduction); and social factors (workforce motivation and safety).

The trajectory of automation is not a linear and homogeneous progress towards fully automated plant. The different weight of the four variables in the firm competitive strategy and the management choice of one priority (for instance efficiency instead of ergonomic problems or flexibility) can partially explain the divergence of firm automation patterns.

Consequently, while the number of automated "functions" or "steps" will increase in the long term, "quality" and "directions" of the implemented automation are less determined. Moreover, each automation trajectory is connected to human resource issues, such as work force motivation, labor costs, safety, individual inertia, management cognitive models, which relationships are not univocal.

First of all, the evolution of lean plants appears to converge to "flexible automation" (Fisher, Jain and Mac Duffie 1994; MacDuffie and Pil 1997). Japanese companies dramatically cut down product variation by 66%, and also launched new models in 1997. These strategies shift competition from static to dynamic flexibility in searching for economies of scope and reduction of changeover costs and times. U.S. automakers still lag far behind their Japanese rivals in changing their factories to produce new models and declare that their objective by the late '90s is to cut this lead-time down to 2-3 months. An example of high investments in automation to improve the flexibility is Nissan. The company developed the Intelligent Body Assembly System (IBAS), a programmable data-base fixture for body assembly, to enable production of any type of vehicle on a single line and reduce tooling of a body main line to a 3 month minimum (Naitoh, Yamamoto, Kodama, Honda 1992). By 1994 Nissan had introduced IBAS body assembly line at all of its domestic plants.

This competitive pressure is undeniable, nevertheless search for flexibility does not lead to the same automation patterns. Some firms, like Honda for instance (EIU 1997), are investing more on product development and value engineering rather than in automation, in order to have fewer platforms for all the range of models. Commonality of model parts is being pursued by Toyota and new models share some major components.

Even if companies invest more on automation, the technological evolution towards flexibility appears to be related to human resources issues. As MacDuffie and Pil (1997) argue, the fully strategic benefits of flexible automation appears to require investments on flexible and skilled workforce. But this is not a well determined choice. Some firms search flexibility but on other hand struggle with workforce constraints (labor offer, work force motivation), therefore try different patterns for automating labor intensive assembly departments, regarding the choice of simplicity (less breakdowns, less maintenance costs) and workforce motivation.

Another sign of convergence is "automation assist", that is development of devices and automation of operations more focused on worker motivation and safety (Fujimoto Matsuo 1993, Fujimoto 1997). The purpose of automation is therefore to assist rather than replace human workforce. Some technological solutions are designed to "balance" customer interests and employee interests (Fujimoto defined this model as lean-on-balance). This automation strategy implies switching to a "worker friendly" plant, so to reduce noise (press shop), work load, psychological stress, etc.. Ergonomic problems and motivational issues are solved in different ways: some firms prefer technological interventions by pushing the automation progress; in others cases retreat from widespread automation is decided partly to improve workforce motivation.

Also unit cost reduction is a common driver of investment in automation. The non-univocal direction of automation strategies is well represented by Toyota. During the late '80s, the company, as well as other carmakers, invested in final assembly automation (Tahara): recent plants (Miyata Kyushu) seem to represent a partial step back driven in part by cost pressures (for instance implying more cautious adoption of robotics), and in part by workers' needs (simple "assisting" equipment to facilitate ergonomics). At Motomachi, where the RAV4 is manufactured, automation was cut back by 66% (comparing to a normal assembly line). To a closer look, the automation degree reduced from 20% to 15% in final assembly and from 90% to 60% in the body assembly line (for example, robots guided by a special camera are replaced by manual installation using a hydraulic-assisted lift in door assembly). Although the RAV4 is not a high volume line, this is symptomatic of the changes taking place. Another example of the automation "retreat" is the Cassino plant, where automation levels were reduced from 100% to 90% in the body shop, and from 25% to 12% in the final assembly. Furthermore, the number of robots was reduced from to 438 to 437 (from 147 to 67 in final assembly), while retooling the Fiat Bravo-Brava model introduction.

Considerations drawn so far lead us to argue that converging patterns coexist with a persistent variety of firm trajectories.

The emerging of specific firm pattern of automation, or eventually plant pattern automation is consistent with different automation degrees measured by the number of robots. Comparing the number of employees per robots and a rough robotics index, such as the number of robots per vehicle per employee cross-plant, differences are evident (even if further considerations could be made using a different ratio).

Table 8.2 Number of robots in selected plants 1994-1996

European and Japanese plants 1994	Body Shop 1994	Painting shop 1994	Final Assembly 1994	Total N. 1994	Vehicles produced 1996	Equivalent workforce 1996	Vehicles per employee	Robot per vehicle per employee	Employees per robot
Fiat Melfi	233	66	36	335	350.000	7.000	50,0	6,7	21
Ford Valencia	382	16	22	420	296.928	5.338	55,6	7,6	13
GM Opel Eisenach	140	10	1	151	161.900	2.390	67,7	2,2	16
Seat - Martorell	-	-	-	302	393.283	8.258	47,6	6,3	27
Renault Flins	-	-	-	300	327.594	7.926	41,3	7,3	26
Nissan Sunderland	255	10	11	276	231.000	3.155	73,2	3,8	11
Mitsubishi - Okazaki	320	40	40	400	120.877	2.000	60,4	6,6	5
Toyota Kyushu Miyata	248			445	126.000	1.910	66,0	6,7	4
Nissan Kyushu n.2 plant *	420			560	187.000	1.700	110,0	5,1	3

* Year 1996

Source: our elaboration on EUROPEAN MOTOR BUSINESS 1995,1996,1997

An important condition for this variety is history. Automation is history dependent (Fujimoto 1993): previous technological investments and organizational choices can "lock-in" the firms to a local optimization. High levels of investment in equipment and cumulative learning of specialized know-how by workers are sunk costs, causing inertia forces against any change leading far from consolidated technical and organizational knowledge. The choice to open green-field plants is illuminating from this standpoint.

At a different level, management cognitive models and past experience are further factors that can account for different firm paths. Particularly, when a firm trajectory is or was very powerful (for example, yield a relevant competitive performance, in reducing labor costs), it might be difficult to switch from one solution to another (for instance, from heavy automation to flexible solutions).

The variety of firm automation trajectories is also fostered by supplier differences and by the fact that most of main carmakers have their own in-house supplier of automation (subsidiary suppliers of robots are for instance Comau-Fiat, Acma-Renault, Mistubishi Heavy Industries- Mistubishi). Fiat-Robogate and Nissan-IBAS are examples of technology developed to satisfy firm's specific needs. These tight relationships tend to deepen the divergence of manufacturing practices at firm level. Nevertheless some companies, like Comau, have multiple customers. Probably the competition within the automation supplier chain can shift the conditions towards the diffusion of common automation solutions in the future (Fine, 1995).

8.4
Convergence and divergence in organizational and IR-HRM practices

Most people studying the auto industry are faced by a fascinating question: is there a process of convergence going on in terms of work organization and HRM practices after the big bang of lean manufacturing? Or, rather, is lean production only an extremely useful metaphor helping scholars and practitioners to map a route, but every firm and every country remain different in the field?

This paper has not the ambition of answering the question (by the way, is there an answer?) and it seeks to articulate a different line of reasoning. In fact, pointing out cross-plant and cross-country similarities and differences is an extremely difficult task. And even when well documented studies are available at national and plant levels (MacDuffie and Pil, 1995), organizational and HRM variables are so strongly intertwined that it is dangerous to single out specific features and make meaningful cross-country and cross-plant comparisons (see Table 2).

Generally speaking, cross-plant differences derive from internal or firm-specific factors (e.g.: performance crisis urging change; possibility to experiment in green field plants; ability to benchmark and learn; workforce age distribution) and from external or country-specific factors (e.g.: labor costs; institutional arrangements; industrial relations and union attitude; labor legislation; customary laws).

Table 8.3. IMVP International Assembly plant data* on HRM and work systems**

Plant Location	CDN	US	JP/NA	JPN	K	FR.SP.IT	UK	N. EUR	AUS	S. AFR	OTHERS
Year	1993	1993	1993	1993	1993	1993	1993	1993	1993	1993	1993
Number of plants in sample	6	19	8	12	6	5	5	11	4	6	4
% of Workforce in Teams	25,2	52,4	67,6	68,3	73,7	40,9	75,7	78,3	66,7	80,3	18
% of suggestion implemented	37,4	41,7	71,9	85,6	43,9	55	24,5	36,9	22,9	14,6	60,7
Extent of job rotation (1=none; 5= frequent, in & across groups)	1,8	2,1	4,1	3,9	3	3,4	3,1	3,8	4	3,3	3,6
Contingent compensation (0=no; 6=extensive)	0,7	1,5	2,1	5,6	2,2	3	2,8	4,5	5	3	1,8
Training for new employees (0=less than 40 hours/year; 3=160+h./y.)	1,5	1,7	2,5	2,7	1,8	3	2,4	2,2	2,5	2	2
Training for experienced employees (0=less than 20 hours/year; 5=80+h./y.)	1,3	1,9	4,3	1,7	2,7	3	2	2,2	2,8	1,8	2,2
Typical size of work team	13,6	12	11,4	13	26,3	13	11,5	14,6	0	17,2	9,2
Team influence: (1=little; 5=much)											
personnel issues	0,6	1,1	1,7	2,5	1,9	2,2	1,2	2,6	2,2	1,6	2,6
work method, problem solving	0,5	1,3	2,6	3,3	2,3	2,4	2,5	2,7	2,9	2,1	2,2

*   means of regional samples
** reproduction kindly authorized by John Paul MacDuffie and Frits Pil

Nonetheless, some considerations can be made from the evidence emerging from international research.

In many cases, recent developments in organizational structures include a decentralization of decision making. Line workers and more generally people involved in the manufacturing process (first line supervisors, technicians etc.) are held responsible for results in terms of quality, cost and delivery.

This increased autonomy is usually associated to a process-based (rather than the traditional function-based) design of plant organizational structure in which each an elementary organizational unit (Kumi at Toyota, UTE at FIAT, UET at Renault, Grupo de Trabajo at Seat, etc.) governs a process (a technological subsystem) and carries out activities such as cost reduction, prevention, variance absorption, quality control, continuous improvement etc.

Units are upstream/downstream related to one another with a supplier-customer like relationship, and they certifies the quality of the operations performed. Interestingly enough, the IMVP-MIT plant survey data shows that the typical size of teams/EOUs is similar across plants and countries (it ranges - average size- from 9 to 26 members but differences are even lower considering that in 7 of the 11 nations or groups, the average team size ranges from 11 to 16 members).

The decentralization of organizational structures at the plant level usually implies a re-design of jobs and roles. For example, the use of management by sight techniques (diagrams, charts, etc.) and on-line information systems with full access to the workers (like at the GM Saturn plant (Whipple, 1993)) change the role of supervisors.

Organizational design, based on processes and elementary organizational units, is usually associated to teamwork, i.e. within each unit there are people working together and cooperating in order to achieve a shared task. However, despite the fact that most, if not all, car makers use this concept, national, regional and firm level specificity shapes it into a number of ways. Murakami (1994) developed an interesting distinction between workgroup and teamwork, and pointed out how the institutional context (for example in terms of work councils, union attitudes, historical background, and nature of the industrial site - green versus brown field) plays a crucial role in differentiating teams across GM-OPEL European plants, even though the Eisenach experience remains the inspiring diffusion principle (EIU, 1995, Haasen 1996). In turn, Eisenach model is itself the result of a learning and diffusion process which combines experiences from other GM plants worldwide (MacDuffie in this book).

The fact that no theoretical definition of team can shape the complex social and cultural environment of a firm should be underlined. To some extent, local specificity and enterprise competencies determines original organizational solutions fitting the industrial context and IR setting (Camuffo and Micelli 1995).

Despite the fact that teams are spreading out in almost all automobile firms, a closer look reveals important differences in terms of functions and objectives to pursue (Fine and Novak, 1996). Team size and composition affect their effectiveness and dynamics. For example, teams that integrate multi-skilled workers with technical competency, such as dedicated line technologists and system controllers, can autonomously govern specific segments of the manufacturing process; but there are also teams composed only by blue collar workers, more or less skilled and properly trained, that try to implement Total Quality Management principles in traditional work contexts and organizational structures.

Team leader's role is also changing. First line supervisor's functions are evolving according to the changing activities assigned to teams. Leaders are in some cases elected and play the blue collar representatives role (Hirschmanian "voice of the workplace" (the case of Mercedes Benz, Western European GM-OPEL plants and some VW plants); instead in other, they are liaisons between firm and worker viewpoints, linking the hierarchical line with the work force (Ford, Fiat, Renault and Seat belong to this group).

Differences in organizational structures are even wider considering the dichotomy between specialization and multiskilling, often considered a fundamental ingredient of lean HRM practices. Cross-plant and cross-country differences are very vast in activating worker "intellectual skill" (Koike 1994) so to improve quality and efficiency. Considering a typical organizational trait related to polyvalence and multiskilling, such as the extent to which job rotation is applied, the IMVP-MIT plant survey data shows that in North-American plants the degree of worker specialization is higher than in Japanese and Korean plants, as well as Japanese transplants in the USA. Similarly, job rotation is less used in North American plants.

Likewise, national research suggests that in Northern European and Australian plants workforce is less specialized compared to the Mediterranean region (French, Spanish and Italian plants), and use of job rotation is more frequent.

On the whole, process focus, decentralization, teamwork, new jobs and multiskilling are important topics to show that human resource centered (rather than automation centered) strategies in manufacturing are homogeneously important across plants and countries.

Nonetheless the effectiveness of human resource centered strategy, i.e. the degree of involvement and contribution of workers, as well as the extent of differences in implementation, is difficult to measure and leads to controversial results.

For example, considering suggestion systems as a proxy of the degree of involvement and cooperation provided by plant workers, the IMVP plant survey data highlights relevant cross-country differences in terms of suggestions per employee. In Canada and USA (non transplant) plants the number of suggestions per employee is comparatively low, as well as the percentage of adoption (38%). The number of suggestions is higher in England (where data should be checked for Japanese transplants) and in the Mediterranean group (French, Spanish, and Italian plants), but remains well below the number of suggestions per employee reported in Japan, Korea and American Japanese transplants, where, in turn, there is a large difference in terms of adoption rates (85% in Japan and less than a half in Korea).

Which is the industrial relations strategy that best supports the implementation of "lean" practices in HR and work organization? Many studies emphasized that lean manufacturing reaches its best results in non-union settings and sometimes suggests it derives from its "mean" nature. Other research, however, keeps a more open approach and advises that different types of industrial relation systems at the firm level are compatible with "lean" HR practices (Rehder, 1994). The obvious reference of this pluralist, non-unilateral approach is the Saturn-GM experience (Bennet, Kochan and Rubenstein 1994) where parallel (union and management) authority lines (partnering) continuously discussed management issues and priorities; apart from North-European plants, where a long standing tradition of cooperative IR constraints ways and modes of lean manufacturing adoption. Also other firms are experiencing new approaches with the unions so to transform the traditional workplace. Other situations are even more complex. For example, Korean car makers have a defensive IR strategy and, to some extent, are overwhelmed by trade union growing strength. As a consequence, they keep pursuing a mere "no-strike" strategy without predicting opportunities related to management/union dialogue (Park and Lee, 1994).

One of the slogans that made the fortune of lean production is that "firms can do the same with 50% less resources" (Womack, Jones and Roos, 1989). From the organizational-HRM perspective, lean manufacturing is achieved through multiskilling, satei (personal assessment), management by sight, job rotation, continuos improvement, etc.. The enhanced flexibility of manufacturing systems derived in part from putting pressure on workers to accept long working hours (Endo, 1994) in physically demanding production systems (Cusumano, 1994).

Pressure impacts positively on reducing waste etc., but can make the system fragile and anorexic (Prahalad and Hamel, 1994). Paradoxically, "lean" efforts can hinder "learn" efforts. In other words a dark side of Kaizen seems to emerge if no slack or redundancy is available (for example when people are pushed to the point of physical and intellectual exhaustion (Fucini and Fucini, 1990)), exposing the system to changes (labor shortage; distance among international plants, etc.) without any protection.

This issue (karoshi) worries both Japanese and Western firms: Japanese firms are facing labor shortage and high turn-over problems (Fujimoto 1994, Berggren 1994; Cusumano 1994); USA and Canadian Japanese transplants have also experienced conflicts regarding lean production related workload intensification (Fucini and Fucini 1990; Unterweger 1992) and could face a labor shortage in the future (Automotive News, Feb.26, 1996). But even European new "lean plants", such as Fiat at Melfi, are facing issues of high demanding work conditions (mid-night and over-night work related to 3-shift operations, long commuting transfer, etc.) which do not fit worker's expectations.

The aim to reduce turnover rate and make production systems less stressful (Fujimoto, 1994) is leading to new emphasis on working conditions and ergonomics, a trend particularly evident now in Japan.

The role of unions is generally important but ends up differently in the diverse contexts. Moreover labor market conditions are also relevant: these issues are more compelling where firms experience labor shortage (Japan, Korea).

Even more important are firm specificity in terms of solution patterns. For example, while in Nissan's main assembly complex, officially called "the factory of dreams", advanced equipment was introduced to raise productivity and eliminate heavy tasks (Berggren 1994), the most recent Toyota plants still work around a conventional assembly chain, but work pace and conditions are changed and made less stressful (Economist Intelligence Unit, 1995).

8.5
The process of diffusion

The diffusion of "lean" IR-HRM practices can be conceived as a dynamic process of inter-firm knowledge creation (Nonaka, 1994). The diffusion process is the spreading out of successful workplace innovations among competing firms. A "form of diffusion" is a system that consists of both: a *media*, which is the channel used to transfer the new knowledge; and a *content* defined as the type of knowledge transferred and its utility for firms acquiring it.

Knowledge diffusion doesn't take place through a simple and natural process of osmosis, thanks to which information gradually passes from organization to organization. On one side, competitive pressures stimulate firms to accelerate this process whenever necessary and possible. On the other hand, social, institutional and cultural constraints are obstacles to an easy transfer of innovations. But also companies, which want to obtain benefits from their competitive innovations as long as possible, intentionally prevent any leakage of information.

The nature of knowledge strongly influences channels of diffusion: codified routines and practices are comparatively easier to transfer and combine than tacit and specific know-how, but they represent only part of the real life of a manufacturing plant. The more tacit organizational and individual knowledge is, the harder it is to spread out very specific traits of complex organizations. Furthermore, the more diverse an innovative principle is from firm's routine, the more difficult it is to recognize and internalize. Moreover, as human resource management and organizational practices are embedded within a specific firm context, the more difficult they are to imitate (Winter 1989).

Despite these difficulties and barriers, firms try to reproduce best practices and innovations of their competitors through different forms of information and knowledge transfer or exchange. Furthermore, since there are comparisons among choices, investments, decisions and techniques (Milgrom and Roberts, 1990), there is an incentive for imitating, adopting or innovating a set of practices rather than single solution (Osterman, 1994).

Historically, the first example of transferring Japanese manufacturing techniques out of Japan were joint-venture forms. The NUMMI and CAMI cases are both well studied examples of interfirm diffusion.

While joint-ventures allow people to work and learn together (through socialization, blue and white collar workers' "experience" a better comprehension of lean management practices), they also represent a costly and risky solution. The complexity of institutional arrangements, cultural factors, industrial relation issues enlarge problems of transferring a complete system of new organizational and technological innovations.

Different and cheaper forms of knowledge exchange and diffusion developed at the end of the '80s. Firms started using different channels to acquire partial solutions to specific problems.

Benchmarking activity for instance is carried out by almost all firms engaged in launching projects of reorganization of   production and organizational systems.

Consultants contribute to building connections between innovative solutions and problem holders. This is a contractual form of know-how diffusion by which organizations can acquire expert codified experiences. Among the various cases, Porsche's collaboration with Shin-Gijutsu   appears to be a good example: Japanese consultants of Shin-Gijutsu (New Technology), previous Toyota executives, helped the firm to implement a Total quality management system. They started in October 1992, invited by Porsche chairman, who wanted to reduce the learning time of the lean production system. The consultants introduced lean practices at Porsche Zuffenhausen factory. Porsche representatives also visited Toyota, Mazda and other Japanese companies to get an insight of the efficient operating practices. By 1995, the management had been reduced by 32% (from six layers to only two), the workforce from 9000 to 6500, and finally the suppliers had been reduced from a high of 900 to 300. As a result the 911 Turbo was built in 40% less time than it took to build the previous model. After the successful experience Porsche set up a production efficiency consulting unit, PES, which now advises other German companies.

University research and world conferences are also "cognitive arenas" where diffusion takes place. On the one hand they work as an interorganizational network, which allow problems and solutions to be discussed from many perspectives, while grouping together highly visible professionals, who contribute to developing and establishing metrics and procedures acknowledged as standards to follow.

But even companies informally exchange information, opening their facilities to their counterparts. It seems to be part of a tacit and long term know-how trading. Toyota for instance, visited by managers from American and European auto makers, took minivan "lessons" from competitors to begin building a minivan in Kentucky in 1997. Toyota's manufacturing engineers, accountants, purchasing people visited minivan plants in the United States and Canada, with a long list of detailed questions provided by the company 1.

Some companies are improving not only the external but the intra-firm learning process. Chrysler's is trying to assemble cars the same way in different plants. It aims at quickly implementing improvements, discovered in one plant, and speeding up the diffusion process. In 1994 at least 1000 workers and managers from the Windsor plant traveled to the St. Louis plant to observe and practice the best operations. The company invested in network connecting among these plants: St. Louis and Windsor have about 50 video communication stations with TV cameras and screens, Graz has 10.

GM-Opel Eisenach plant set a standard and a set of organizing principles which guided restructuring of other European GM-Opel plants, like Luton and Bochum (EIU, 1995)

The role of individuals (managers and workers) is usually underestimated in the diffusion process. Instead, individuals play a twofold crucial role in the inter-firm "contamination process" of different knowledge creation cycles.

Firstly, individual social networks facilitate linkage of information and know-how between different organizations. Occasional contacts between professionals or employees of different companies can be channels of informal know-how trading (Von Hippel 1987). Personal contacts, phone calls etc. are used to exchange ideas, and ask or give suggestions on specific technological or organizational problems. The vice president of manufacturing at the Toyota Georgetown Kentucky plant recently said, in an interview, that there is nothing unusual about asking competitors for technical help, and that he often talks with his counterparts in other companies.

Secondly, personnel turnover (especially management) within and between firms allows companies to spread and increase experience. For example, Japanese firms intensively used their manager mobility from Japan to the USA as a diffusion form of core traits of their production system in overseas operations. It is the case of the Toyota manager who presided over the RAV4 line, and was chosen to run the Toyota plant in Georgetown Kentucky. The above mentioned Toyota Georgetown manger was a former GM manager. The president of BMW Manufacturing Corp. came to his present job after 16 years at Honda of America Manufacturing. Another example comes from Europe: in the early 1990s, SEAT Martorell plant's director had formerly been at Nissan's plant in Sunderland, joining the Spanish operation with a team of UK colleagues (EIU, 1993). This form of knowledge transfer is not risk-free. Problems often arise especially when different organizational cultures collide.

8.6
European peculiarities

The process of diffusion of "lean manufacturing" touched Europe the late 1980s and still has to produce its effects. From this standpoint, European car makers can be considered late adopters of LP. This delay has some negative implications, such as European plants are still far from productivity and quality world class performance on average, despite a good catching up in recent years (MacDuffie and Pil, 1995).

Nevertheless, being late adopters probably had some advantages, for instance the possibility to learn from US Japanese transplants, i.e. from Western derived and digested applications of LP. This also allowed, at least in some cases, the avoidance of non effective solutions and mistakes made elsewhere.

On the whole, LP diffusion in Europe seems to be more differentiated than in the US. This probably comes from the relevant role of national institutional factors, that tend to shape the adoption patterns (e.g. labor legislation, labor costs, labor relations systems). As a consequence, clusters of modified versions of LP can be identified (UK JPN transplants; "Mediterranean LP") (Camuffo and Micelli, 1995), although it is difficult for Europe to see if company factors are more relevant than country factors in shaping differences (MacDuffie, 1995).

Plant performance is affected by organizational, IR-HRM variables, but not in a deterministic and univocal way. Instead, as Figure 3 shows, there are multiple possible combinations of IR-HRM variables, depending upon firm-specific and context-specific factors, that lead, together with other (technological, etc.) variables, to a given level of productivity and efficiency.

ORGANIZATIONAL AND HRM-IR IMPACT ON PLANT PERFORMANCE

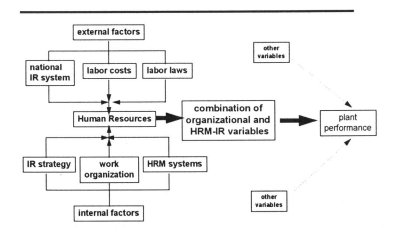

Figure 8.1 How organizational and IR-HRM variables impact plant performance

Table 4 reports some data for selected European plant. Performance levels, in terms of vehicles per employee according to the Economist Intelligence Unit measurement, are different. Interestingly enough, the values for two simple IR-HRM variables like labor costs and number of shifts operated by the plant, are different and greatly impact, though in different ways and from different perspectives, on plant efficiency.

Table 8.4 Productivity, labor costs and work shifts in selected European plants

Plant	Vehicles per employee 1996	Vehicles per employee 1995	Vehicles per employee 1994	Typical annual salary (ECUs)	Number of work shifts
GM-Opel Eisenach	67.7	71.9	59.3	22,000	3
FIAT Melfi	50.0	64.3	54.5	12,500	3
NISSAN Sunderland	73.2	56.7	54.0	17,500	2
HONDa Swindon	64.2	55.9	n.a.	15,500	2
GM-Opel Zaragoza	54.2	54.0	51.0	15,800	3
FORD Valencia	55.6	52.9	42.0	15,800	2
TOYOTA Burnaston	66.9	52.1	n.a.	17,500	2
RENAULT Flins	41.3	46.9	48.0	17,500	3
SEAT Martorell	47.6	43.7	35.1	15,800	3

Source: Economist Intelligence Unit, 1995,1996, 1997; authors's elaboration

Moreover, the impact of different IR-HRM variables on plant performance is difficult to assess.

Labor costs widely differ across Europe. These variances reflect differences in wages, often related to skill levels, with similar compensation (social security etc.) or in labor conditions (working time, holidays, etc.). The impact of labor costs on plant performance could not be overemphasized, since a 10% difference in hours per vehicle or vehicles per employee is hardly comparable between plants having a 50% difference of hourly labor costs.

The workforce age distribution and the skill endowment of a plant is another major variable. It is interesting to note that, at least according to the data in Figure 4, some of the top performing European plants are greenfields where workforce is relatively young, newly hired and appropriately selected.

Strikes and absenteeism are other major factors. The former largely depends on the unions (if it is a unionized plant) and hence it is largely affected by the company and national IR systems. It is also interesting to note that some top performing European plants are non-union, and some have a cooperative, partnership based IR system.

Regarding work organization, the process based approach of plant organizational structures (though with differences in size and composition of EOUs), the role design (Murakami, 1994; Camuffo and Micelli, 1995)., and the number of organizational levels (trend to flat organizational structures in most top performing European plants) have already been mentioned.

The operation modes of the plant in term of work shifts, despite the appearance (2 or 3 shifts), entail wide variations. Typically, for example the UK Japanese plants work on 2 shifts, very different from typical European shifts. Furthermore, there are differences also in 3 shifts operations. For example the Fiat Melfi plant works on 3 shifts, 6 days a week, while GM, Renault, Rover and SEAT plants usually operate a conventional 3 shift.

Skill endowment and level of investment in training both new hires and normal employees are factors impacting performance, as well as the overall workforce attitude towards change, motivation and commitment. Plenty of literature is now available on how auto makers strive to increase motivation, commitment and attitude towards changes (contingent compensation, QC and suggestion systems, Kaizen activities, etc.).

However, these efforts are not always successful and, most of all, are not robust and permanent over time. Cultural factors often hinder these initiatives. In turn it must be underlined how the implementation and success of the above mentioned systems seem much easier in different contexts (e.g. emerging countries new plants) where cultural flexibility is greater and a certain degree of naïveté allows a easier rooting of this organizational approach. This is, for instance, the experience of FIAT. While the Fabbrica Integrata concept was implemented successfully at the Melfi plant and at the Brazialian Betim plant, it met more difficulties and resistance in the older Italian plants, but also in the Cordoba Argentina plant, where Fiat experienced labor relations problems in implementing the production of Palio, its worldcar.

8.7
Balancing lean production and lean employment

European car makers are striving to survive. The adoption of lean manufacturing has taken place throughout the 1990s in a time of recession and troubles. European markets are mature. There is overcapacity in almost every country. Price battles (for instance Ford theorized it as "price destruction theory") put pressure on margins and emphasized efficiency search. As a consequence, especially in countries with high labor costs, scholars and practitioners are wondering if production will move outside Europe sooner or later. Recently, for example, the President of German VDA said that high labor rates are taking the auto industry away from Germany. With wages 25% higher than Japan and still rising, Germany is losing its attraction as a location for the auto industry, resulting in widespread redundancies (Automotive International, 26, March 1996).

Table 8.5 Comparison of Labor rates (DM per hour, 1995)

Country	1980	1985	1990	1995	1995/1980 (%)	1995/1990 (%)
Germany	28.01	35.04	43.95	60.02	114	37
Belgium	28.14	30.67	33.49	40.07	42	20
Sweden	28.60	36.28	42.78	36.16	26	-15
NL	23.38	28.62	29.35	33.70	44	14
France	19.66	25.14	26.09	30.11	53	15
Spain	12.63	19.91	27.01	26.39	108	-2
UK	14.95	21.41	25.05	25.33	69	1
Italy	17.73	25.63	28.63	25.32	43	-12
USA	24.83	55.75	33.13	35.69	44	8
Japan	13.26	25.23	26.83	45.81	245	70

Source: VDA

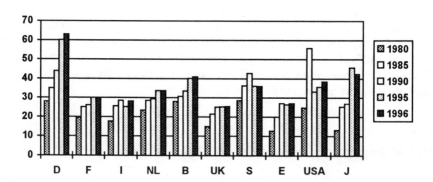

Source: VDA

Figure 8.2 Comparison of Labor rates (DM per hour, 1997)

Therefore, labor represents, for European car makers, a major issue. In fact, two contradictory needs must be balanced: the search for competitiveness driving adoption of lean manufacturing and the search for a socially accepted industrial adjustment driven by the "Lean employment" trend (Auer and Speckesser, 1996).

Lean employment in the European Auto industry consist of:

- a systematic shrinking of employment levels, although with different timing and rates, in all European countries;
- a decrease of assemblers' workforce
- a relative increase in suppliers' proportionate share of employment (mostly as a result of outsourcing)
- a decline (at least in some countries) of working hours and union density

The Eurostat data of Figure 6 clearly shows the trend towards "lean employment" in the auto industry.

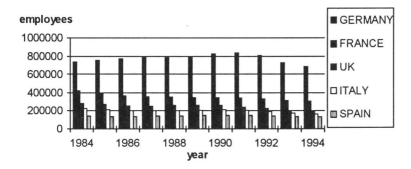

Source: Eurostat, 1995

Figure 8.3 Employment decreases, <u>at different</u> times and rates, in the UK, Germany and France

Structural labor shedding, given the importance of the auto industry in Europe, represents a major challenge. Auto manufacturers and national governments have to design complex strategies of industrial adjustment.

Technological change is a "fundamental force" in shaping the adjustment pattern of economy, industry and firms (Freeman 1988). Several relevant traits of technological change should be taken into account once one wants to analyse automation impact on automotive industry and automakers adjustment strategy. Firstly, technological progress "generally exhibits strong irreversibility features" (Dosi 1988), and the long term mechanisation process of production exhibits an irreversible labour saving impact (see Volpato in this book). Secondly, the automation progress and its work force reduction effect at firm level depends on technological innovation strategies along the entire added value chain of the firm (product development, logistic, distribution process). Finally, automation demonstrates a *schumpeterian* result of destructive innovation. The *quantitative* impact of job destruction is the most evident (and controversial) effect but it is not the only one. Automation has also a *qualitative* effect in terms of organizational change. The substitution of manual operations with automatic ones changes the ways of production. For instance it requires HRM changes towards flexible employment (temporary workers) and flexible work schedules.

These strategies largely depend on enterprise governance modes (company stakeholders, role of workers and unions etc.), competitive strategies (product/market/technology combinations) and core competencies (history matters).

Moreover, they are shaped by the institutional setting which affect the feasibility and cost-effectiveness of a given strategy for a given firm.

Generally speaking, these strategies could be based on internal and/or external flexibility (Auer and Speckesser, 1996).

An *internal flexibility based strategy of adjustment* is typically based on:
- the maneuver of the internal labor market by means of mobility (e.g. of workers across plants), training and relocation;
- a strong commitment to keep employment level and job security pledge;
- the design of flexible working hours and schedules
- the adoption of flexible compensation schemes.

Instead, an *external flexibility based strategy of adjustment* is typically based on:
- the maneuver of the external labor market by means of flexible recruiting (apprenticeship, temps, etc.) and outplacement (early retirements, state financed redundancy, fund etc.);
- a speed-up of generational turnover, what typically happens is the company hires and downsizes at the same time;
- a core-periphery internal labor market model where a core of young, skilled workers co-exist with a buffer of "marginal" employees.

8.8
Volkswagen, FIAT and Volvo: comparing adjustment strategies

European patterns of adjustment to "lean employment" are different. This largely stems on the one hand from the firm's heritage and on the other from the national institutional setting.

In order to show how these factors impact on the adjustment patterns, let us compare the Volkswagen, FIAT and Volvo cases. As highlighted in Figure 7, two companies (VW and Fiat) faced, in the early 1990s, a performance crisis that urged restructuring and downsizing in order to increase competitiveness. After the early 90s crisis (Berggren 1992) and the failed merger with Renault, planned to gain a more competitive and global dimension, Volvo focused on a niche strategy and on flexibility to increase its competitiveness. The firms used the LP concept as a reference with different approaches. Nevertheless, the internal and national premise were completely different as far as the national system, labor costs and wages, modes of skill formation and development, enterprise governance, industrial relations system are concerned.

Table 8.6   Strategies for adjustment:VW, FIAT and VOLVO (1993-1995)

Premises	VOLKSWAGEN	FIAT	VOLVO
National system	Standort Deutschland	Unfavorable	Neo-corporatist
Labor cost	High	High; low in South Italy	High
Wages	High	Low	High
Workforce age distribution	Relatively balanced	Relatively skewed	Relatively balanced
Skill formation and development	Based on German vocational, dual training system	Based on internal training (incentives to training contracts -Cfl)	Based on Sweden vocational -
Enterprise Governance	Mitbestimmung (information, codetermination)	Managerial Unilateralism Confrontative Industrial relations	Union supported innovation strategy
Union "voice" medium	Betriebsraet	R.S.U.	Workplace clubs
Declared workforce redundancies	30.000	17.000	n.a.

Source: Camuffo and Volpato, 1995; Hartz, 1994 and 1996; MacDuffie 1995

The adjustment strategies designed and implemented by the three firms in the 1993-1995 period were rather diverse.

Volkswagen's strategy was designed and implemented in 1993 (union agreement) and then developed in 1995 (Hartz, 1994 and 1996). It consisted of:
- 4-day-week -Viertagewoche- (reduction of weekly working hours to a flexible average of 28.8 with the possibility to increase it according to the market needs) entailing also compensation adjustment;
- Training -Blockzeit- (possibility to spend 3-6 months a year in training)
- "Relay" between older and younger workers -Stafette- (slow ramp-up of working hours for new hires while working time of older workers accordingly decreases).

Fiat's strategy was designed and implemented in 1993 (union agreement in 1994) after a number of other plant-level innovative union-management agreements accord (Camuffo and Volpato, 1995). It consisted of:

- Ordinary state redundancy fund -Cassa Integrazione Guadagni Ordinaria- (52 weeks in two years) in order to absorb temporary demand downturns;
- Special redundancy fund -Cassa Integrazione Straordinaria- granted by the State contingent on the declaration of the firm's crisis status;
- Early retirements -Mobilità lunga-;
- Working hours reductions -Contratti di solidarietà- implying also a reduction of wages;
- Mobility within the firms of the Fiat group.

Volvo's competitive leverages, as a low volume automaker (500.000 vehicle planned by 2000), are high product quality3 and product innovation. The introduction of new models in order to widen the range was main objective of the company and led Volvo to invest on product development and search flexibility at production level.

Volvo's recent adjustment strategy can be analyzed considering the early experience of Uddevalla plant, re-opened in 1996[2] and run by a new subsidiary "Autonova" (joint venture of TWR 51% and Volvo 49%), It seems like a step towards a more "lean" work system in search of workforce flexibility and cost reduction:

- brownfield - 60% of the employees are part of the previous workforce
- large and polyvalent teams- integration of assembly workers and material handling ones, but less holistic competency than previous experience (a more simple part of the process to remember even if still wide, seems to help people's capacity of problem solving under time pressure)
- systematic job rotation
- 35% of Uddevalla's workers are women
- industrial relations for employees flexibility - only one union agreement covering all the employees (white and blue collars and the engineers) previously represented by different occupational unions in Volvo plants. This is a first experience in Sweden as well. The objective of this agreement is to increase polyvalence among different jobs
- absence of the union "voice" at plant level
- flexible hours - management can raise 10% the normal hours to increase production
- training commitment required to work force - 100 hours per year
- flexible wage planned - related to objectives of quality and productivity
- entry wage lower than other Volvo plants but possibility to raise it to a higher level

---

[2] Production of the C70 coupe began at Uddevalla during spring 1997 with initial rate of 7000which will gradually increase to 20.000 car per year

On the whole, both VW and FIAT are mixed strategies based in part on internal and in part on external flexibility. Fiat's strategy is more external flexibility based while VW's strategy is more internal flexibility based.

These features relate to the major strategic thrust for both companies to lower B.E.P. But while Fiat aimed at it directly through rightsizing, in order to speed up the generational turnover and enhance its flexibility, ability to learn and to adapt, VW aimed at it through the principle of "the firm breathing with the client". Employment can be maintained (at least in principle) if the workforce and organization are capable to flexibly adapt to the pace of the market. And this can be achieved maneuvering working time and, in part, compensation.

The VW strategy is highly acceptable from the standpoint of social consensus since it limits headcount reduction (only one third of the 30000 redundancies). However, if "lean employment" remains a structural trend, it is not likely to be cost effective and robust. Besides, this strategy was possible because of high wages and peculiar skills/training system in Germany. In fact, on one hand workers tend to not be sensitive to compensation flexibility if wages are relatively high; on the other hand, the high skill levels of the workforce is as an incentive to keep employment levels up . Finally, the strategy was coherent with a time responsiveness based strategy (the "2-2-2" VW principle).

The Fiat strategy is aimed at speeding up generational turnover in order to correct a skewed workforce age distribution. It is cost-effective, also because it takes advantage of institutional arrangements facilitating outplacement and new hiring (and South Italy lower labor costs). It is coherent with the ongoing union demise. However, it could harm internal climate and result in labor conflicts (especially if older plants close).

Finally Volvo's choice can be explained considering both the economic and institutional factors: Company's good market performance during the 90s (from 1991 to 1995 sales increased 37.022 to 83.340 and Volvo Car Corporation hired 2000 new employees in 1996) and the small dimension of the plant (about 20.000 vehicle) can allow management to work out Volvo's reflective production system (Ellegard 1997). On the other hand, the company's not so good profit performance and competition in international markets (main Volvo market) forced the firm to look for solutions increasing system efficiency.

8.9
Conclusion

The paper started by pointing out that the task of analyzing implications of "lean" IR-HRM practices diffusion will be long and demanding, comparable with the work of the "agenda" of Scientific Management during this century.

Then based on the field evidence of IMVP and other research, the paper analyzed the diffusion process of LP, highlighting how differences across plants remain partly due to company factors (history matters) and partly to national, external conditions inducing diverse responses to competitive challenges , while competitive pressure moves firms to converge to similar inspiring concepts.

From this analysis three issues can be identified for further discussion and research on the dynamic of LP diffusion:

- diffusion process is facilitated by research, benchmarking and discussion in appropriate arenas.
- adoption is a discovery process, that does not result in an unique set of solutions. It proceeds by trial and error, and requires a long and costly process of search, learning and organizational change.
- diffusion is a combination of internal and external learning processes.

The European car industry is peculiar, since LP diffusion is accompanied by a general strive to survive and "lean employment".

This situation challenges European car makers, governments and unions and asks for adjustment strategies which must balance the search for efficiency and competitiveness with socially acceptable ways of absorbing redundant workforce. Firm-level innovation and institutional change, involving all the relevant actors, seem to be, even in this field, the only appropriate response.

8.10
References

Auer P., Speckesser S., 1996, "Labour markets and Organizational change. Future working structures for an ageing workforce, WZB working paper.

Berggren C., 1994, "Japan as number two: competitive problems and future of alliance capitalism after the burst of the bubble boom", Work, Employment & Society, Vol.9, n.1, pp.53-95.

Camuffo A., Micelli S., 1995, "The mediterranean lean production", Paper presented at the 10th IIRA World Congress, Washington DC.

Camuffo A., Volpato G., 1995, "The labour relations heritage and lean manufacturing at FIAT, The International Journal of Human Resource Management, Vol. 6, No.4.

Cerruti, G. 1995, "La razionalizzazione alla FIAT Auto: dalla crisi del taylorismo ai dualismi della lean production", Quaderni di ricerca, IRES, n.17.

Cohen W.M., Levinthal D., 1990, "Absorptive capacity: new perspective on learning and innovation", Administratuive Science Quaterly, pp.128-152.

Coriat, 1991, Penser à l'envers, Christian Bourgois, Paris.

Cusumano M.A., 1994, "The limits of "Lean", Sloan Management Review, Summer.

Dosi G., 1988, The nature of the innovative progress, in Dosi G., Freeman C., Nelson G., Soete L., 1988

E.I.U. (The Economist Intelligent Unit), 1993, "Inside SEAT's Martorell plant" Motor Business Europe, second quarter.

E.I.U. (The Economist Intelligent Unit), 1995, "Europe's leading car plant: comparative productivity audit" Motor Business Europe, 3rd quarter.

E.I.U. (The Economist Intelligent Unit), 1995, "Inside Fiat's Melfi plant: rivalling japanese productivity?" Motor Business Europe, second quarter.

E.I.U. (The Economist Intelligent Unit), 1995, "Inside GM Vauxhall's Luton plant: brought into the 1990s with Japanese methods" Motor Business Europe, 4th quarter.

E.I.U. (The Economist Intelligent Unit), 1995, "Inside renault Flin's plant:setting the pace for minicar production" Motor Business Europe, 1st quarter.

E.I.U. (The Economist Intelligent Unit), 1996, "Europe's leading car plant: comparative productivity audit" Motor Business Europe, 3rd quarter.

E.I.U. (The Economist Intelligent Unit), 1996, "Inside Ford Valencia's plant: home of the Ka" Motor Business Europe, 1st quarter.

E.I.U. (The Economist Intelligent Unit), 1996, "Inside Honda's UK transplant: ripe for expansion" Motor Business Europe, second quarter.

EIU, 1996, Motor Business Europe, Europe's leading plants: comparative productivity audit:, 3th quarter

EIU, 1997a, Motor Business Japan, Productivity audit: Japan's leading plants, 4th quarter

Endo, 1994, "Satei (Personal Assesment) and Interworker Competition in Japanese Firms", Industrial Relations, Vol.33, n.1

Dosi G., Freeman C., Nelson G., Soete L., 1988, Technical change and economy theory, Pinter Publishers, London

Freeman C., 1988, Introduction, in Dosi G., Freeman C., Nelson G., Soete L., 1988

Fucini J. J. and Fucini S., 1990, Working for the japanese, The Free Press, New York

Fujimoto T., 1994, "The limits of Lean Production.On the Future of the Japanese Automotive Industry", International Political Gesselschaft, n.1

Garrahan P., Stewart P., 1992, The Nissan enigma. Flexibility at work in a local economy, Mansell, London.

Haasen A., 1996, "Opel Eisenach GMBH- Creating a high productivity workplace", Organizational Dynamics, spring.

Hamel G., Prahalad C.K., 1994, Competing for the future, Harvard Business School Press, Cambridge.

Hartz P., 1994, Jeder arbeitsplatz hat ein gesicht. Die Volkswagen loesung, Campus, Frankfurt.

Hartz P., 1996, The company that breathes: every job has a customer, Springer Verlag.

Ittner C.D., MacDuffie J.P. (1994), Exploring the Sources of International Differences in Manufacturing Overhead, IMVP Research Briefing Meeting Working paper, June

Kenney M. and Florida R., 1993, Beyond tmass production, Oxford University Press, New York.

Kenney M., Florida R., 1991, "Transplanted organisations: the transfer of Japanese industrial organisation to the U.S.", American Sociological Review, vol.56, pp.339, 398

Kochan T., Lansbury R., 1996, Changing employment relations and governance in the International auto industry, IMVP working paper.

Kochan T.A., Bennett M., Rubinstein S., 1994, "The Saturn partnership: Co-management and the re-invention of local unions", in Kaufman B., Kleiner M., Employee representation: alternatives and future directions, Madison, Wisconsin, IIRA.

Koike K., 1994, Learning and incentive systems in japanese industry, in Aoki and Dore 1994.

MacDuffie J.P., 1995, International trends in work organizations in the auto industry: national-level vs. company-level perspectives", forthcoming in IRRA 1995 Research Volume.

MacDuffie J.P., Pil F., 1996, "From fixed to flexible: automation and work organization trends from the International assembly plant survey, IMVP working paper.

MacDuffie J.P., Pil F., 1996, "Performance findings of the International assembly plant survey, IMVP working paper.

Milgrom P.J., Roberts, J., 1990, "The evolution of modern manufacturing: technology, strategy and organization", American Economic Review, vol.80, n.3.

Mueller F., 1994, "Teams between Hierarchy and Commitment: Change Strategies and the Internal Environment", Journal of Management Studies, 31:3, may.

Murakami, T., 1995: "Teamwork & Participation in the German car industry" IMVP-IR Working Paper, presented at the 10th IIRA World Congress, Washington DC.

Nonaka I., 1994, "A Dynamic Theory of Organisational Knowledge Creation", Organisation Science, Vol.5, No.1.

Osterman P., "How common is workplace transformation and who adopts it?", Industrial and labor relations review,, vol.47, n.2, february.

Park, Y. and Lee H., 1994, Industrial relations and human resource practices in the korean automotive industry: recent development and policy options, Korean Labor Institute WP.

Rehder W., 1994, "Saturn, Uddevalla, and the Japanese lean system. paradoxical prototypes for the twenty-first century", The International Journal of Human Resource Management, Vol.5, n.1, february.

Robertson D., Rinehart J., Huxley C., 1992, "Team concept and Kaizen: japanese production management in a unionized Canadian auto plant", Studies in Political Economy, 39, Autumn

Robertson, P., and others, 1993, "Team concept and kaizen: japanese production management in an unionized canadian auto plant", Studies in Political Economy, Autumn

Shimokawa, K., Jurgens, U. and Fujimoto, T. (eds), 1997, Transforming Automobile Assembly. Experiences in Automation and Work Organization , Springer, Berlin

Unterweger P., 1992, Lean production: Myth and reality, IMF Automotive Departmente

Von Hippel E., 1987, "Cooperating between rivals: informal know-how trading", Management Science, vol.37, n.7-791-805.

Winter S., 1987, "Knowledge and competencies as strategic assets"; in Teece D.J. (ed.), The Competitive Challange, Mc Graw Hill, New York

Womack J.P., Jones, D.T. and Roos D., 1990, The Machine That Changed the World, Rawson Associates, New York.

Womack J.P., Jones, D.T., 1996, Lean thinking. Banish waste and create wealth in your corporation, Simon & Schuster, London.

9   The Transfer of Organizing Principles in the World Auto Industry: Cross-Cultural Influences on Replication at Opel Eisenach

John Paul MacDuffie

The world auto industry is currently experiencing the gradual diffusion of new organizing principles for manufacturing work, linked to a model of the overall production system -- often called "lean production" (Womack, Jones, and Roos, 1990) or "flexible production" (MacDuffie, 1995) -- that has significant productivity and quality advantages over traditional mass production. These organizing principles are increasingly likely to become the dominant model for automotive manufacturing. This provides an important opportunity to assess the extent and nature of the diffusion process, particularly across national borders, in light of past theories about the transfer of this kind of knowledge.

Historically, research on the transfer of principles of organizing work suggests that there is substantial national variation in how certain dominant principles are understood and applied, and that such variation is historically persistent (Kogut and Zander, 1992). For example, the overlapping principles of Taylorism and Fordist mass production that diffused broadly in the U.S. were also much studied by companies in Europe and Japan. However, the adoption of those principles in these other countries was slow, partial, and affected by cultural influences at both national and company levels. In Europe, the persistence of craft traditions was the dominant cultural influence on the adoption of Taylorist work organization (Lewchuk, 1988), and in Japan, constraints on the implementation of Fordist mass production led to the development of the alternative organizing principles associated with lean or flexible production (Cusumano, 1985).

But while past patterns of work organization show persistent **differences across** different countries and **convergence within** countries and companies, I will argue that the emergence of a new set of dominant organizing principles in the auto industry has created the conditions for more **convergence across** countries and **divergence within** countries and even companies, as follows. The development of "flexible " or "lean" production principles for organizing work led to divergence in country economic performance in automotive manufacturing. The transfer of these principles by Japanese companies to their own overseas plants in the U.S. and Europe provided convincing evidence that the principles could be effective and durable in other cultural contexts (Florida and Kenney, 1991). This, plus the competitive pressures associated with the increasing internationalization of automotive product markets, has prompted efforts by U.S., European, and Korean companies to adopt those principles in both their home country plants and in their overseas operations.

Furthermore, the new organizing principles are increasingly legitimized in the institutional environments surrounding company strategic decisions -- the business press, Wall Street, the actions of "leading" companies. Thus, from the perspective of organizational change, the current period in the auto industry may represent a time of "punctuated equilibrium" (Tushman and Romanelli, 1985) -- the interruption of a long period of stasis for a dominant model, during which new technologies and organizing principles begin to take hold -- and of institutionalization pressures favoring convergence towards the new model based on some consensus that it represents "best practice" (DiMaggio and Powell, 1993).

New data collected from an international sample of assembly plants suggests that the diffusion of these principles is occurring and that performance across countries is now converging (MacDuffie and Pil, 1995). Nevertheless, it is also clear that these principles are adapted to local needs and conditions as they are transferred, both intentionally and unintentionally (i.e. because the principles or their implementation are not well understood). Even where simple imitation is the goal, there is considerable adaptation and innovation (Westney, 1987). There is great variation in how different companies are approaching the adoption of flexible or lean production, both in the importance they ascribe to this shift in manufacturing strategy and in their approach to implementation.

One increasing common approach for companies trying to transfer lean or flexible production systems is to open a new, "greenfield" plant and to attempt to replicate closely the production system of another, successful plant. Replication as a transfer strategy can be differentiated from other approaches in various ways (Winter, 1995; Nelson and Winter, 1982). Replication requires extensive access to the "learning model" or "template" (Kogut, 1993) for the change effort, because it assumes that new organizing principles are best learned experientially rather than didactically. It emphasizes the value of limiting the sources of variation in a new operation, by duplicating features of the template (e.g. product, process flow, technology, work organization and methods) as much as possible during the start-up period. It reflects the belief that policies and practices "not invented here" are legitimate to transfer in their entirety, rather than attempting to "reinvent" all procedures to establish a sense of local ownership (Szulanski, 1994).

This paper will analyze one prominent replication effort: the Eisenach plant of General Motors in Europe. Eisenach is a new, "greenfield" plant established in the former East Germany with the expressed intent of being the "lean production" exemplar for GM in Europe -- much as the NUMMI joint venture in the U.S. between General Motors and Toyota was for U.S. auto companies in the mid-to-late 1980s. The managers and "advisors" implementing lean production at Eisenach are mostly Americans with prior experience at NUMMI and also at another joint venture plant in Canada (called CAMI) between GM and Suzuki. They bring these ideas to a company long dominated by German ideas of design and manufacturing (it was once Opel) and to a workforce whose notions of work were shaped under a centrally-controlled socialist economy. Thus Eisenach is a particularly fascinating place to explore how new organizing principles that are highly legitimized by the external environment and the parent company are introduced into a very different cultural context.

Building on much recent work on the dynamics of cross-cultural interaction in an organizational context, I will argue that the use of modal cultural tendencies as the analytical lens is far too restrictive. There is rarely a direct mapping of national cultures onto organizational cultures (Hofstede et al, 1993; Brannen and Salk, 1994). Indeed, managers who are "natives" in terms of a national or corporate culture may deliberately act or view situations from a "marginal" rather than a "normal" perspective for that culture (Weiss, 1994). Furthermore, managers who have accumulated diverse experiences during their careers, particularly in multinational firms, have multiple national and organizational cultures that they can draw upon for ideas and as models for behavior (Brannen and Salk, 1994) -- what could be considered a cultural repertoire. Managerial perceptions -- what they see as important in a given situation -- and actions -- what behaviors they select from their cultural repertoire -- will be affected by their past experiences in various cultures, together with what cultural norms are legitimized in their current setting.

9.1
GM Europe's Eisenach Assembly Plant Background

Most of the material in this case is based on one initial period of field work at the Eisenach plant -- three days in May 1994 -- and subsequent interviews with other GM Europe managers, both at corporate and plant levels. Further field work at Eisenach is planned for the fall of 1995. My initial visit to Eisenach involved extensive plant tours and interviews with the plant manager and the seven members of the top management team, as well as six of the American and Canadian advisors who were paired with the management team. My previous fieldwork at NUMMI and other Japanese transplants in North America also informs my observations of the replication effort at Eisenach.

The story of how the Eisenach plant was established reveals how rapidly the competitive terrain for the European auto industry has shifted in the last six years. The president of GM Europe, Louis Hughes, was already thinking about the prospects for a new plant in Eastern Europe when the Berlin Wall fell in November 1989. By January 1990, he had initiated joint venture talks with Automobilwerk Eisenach (AWE), a state-owned East German firm making a vehicle known as the Wartburg -- a "high end" product, in relative terms, amid the limited array of cars (Trabant, Skoda, Lada) available in Eastern Europe.

By March 1990, a joint venture was in place, with initial plans to build Opel Vectras from knock-down kits at the old Wartburg plant. The production of Vectras began just seven months later, in October, and gave the first team of GM Europe managers their first chance to assess the East German workforce. But from the start, this transitional phase was intended to be brief. By February 1991, groundbreaking for a new Opel plant in Eisenach took place, intended for production of Opel's Corsa and Astra models (subcompact and compact sizes, respectively) at a planned capacity of 150,000 per year. Since these products were also being built at GM's plant in Zaragoza, Spain, many of the components could be sourced from existing suppliers and sent to Eisenach (mostly by train, given the limited and already congested highway infrastructure in East Germany). Zaragoza was also assigned as the "lead" plant for Eisenach -- the source of product-specific technical information about the production process -- although not as the source of production system ideas (as described below).

By April 1992, Vectra production ended in the old plant and preparation began for the startup of Astra production in the new plant. Between April and September 1992, when the first Astra was built, workers from the old AWE facility were being screened and selected to form the new workforce, and then trained at the new site. (I will discuss the dynamics of this hiring process in more detail below.) Astra production ramped up slowly over the next eight months, with a second shift added in May 1993 and the production of Corsas beginning in June 1993. By October 1993, a third shift of workers was added and the plant reached full capacity by December 1993.

Eisenach was initially planned by Opel engineers as a plant that would resemble other Opel plants in its overall production system, albeit with newer technology.[1] But as the plant was being built, Lou Hughes decided that Eisenach should become GM Europe's "learning laboratory" for lean production. Hughes,

---

1   This initial plan is most evident in the physical design of the Eisenach plant. Most of the Japanese transplants in North America have a compact, space-efficient plant with all operations under a single roof. Even NUMMI, which occupies an old GM plant, rearranged its production layout to use only a portion of the existing facilities. But Eisenach, like many other newer U.S. and European plants, has three separate buildings for welding, paint, and assembly and each building is generously sized. Lengthy conveyors between the buildings guarantee a higher level of in-process inventory than many lean production plants see as desirable. The distance between plants also makes cross-department communication more difficult.

along with current GM CEO Jack Smith, had been involved in the negotiations for the GM-Toyota joint venture (known as NUMMI, for New United Motors Manufacturing Inc.) in the early 1980s and, more than most GM executives, had absorbed the crucial lesson of NUMMI's startling advantages in productivity and quality over GM's more traditional plants -- a different production system, based on different principles and with a different approach to managing both people and technology. Given the widespread perception that GM in North America had not been able to learn effectively from NUMMI, Hughes was determined to make Eisenach a more successful experiment -- not just as a stand-alone plant but, more importantly, as a training ground and launch pad for the diffusion of lean production within GM Europe.

9.2
Key Players

Before discussing the transfer and implementation of lean production concepts at Eisenach, it is important to know something about the key players: the plant manager and his management team, the young advisors from North America, the engineers at Opel headquarters who had responsibility for the plant's technical design, the East German workers from the old AWE plant, and the leadership of the union/works council that was established at Eisenach.

Plant manager and management team. Chosen as the first plant manager was Tom LaSorda, an American whose most recent experience had been as manager of the GM-Suzuki joint venture known as CAMI (Canadian Automotive Manufacturing Inc.) in Ingersoll, Ontario which opened in 1989. The CAMI plant can be characterized as a "lean production" plant but it differs in various ways from NUMMI, given the influence of Suzuki as a joint venture partner vs. Toyota. (These differences are described in more detail below.) Its launch under LaSorda was viewed as very successful so, despite the fact that it was only two years since CAMI had opened, LaSorda was picked to oversee the launch at Eisenach.[2]

---

2   Since LaSorda's departure, CAMI has had a more troubled period, marked by declining worker satisfaction, some quality problems, and considerable conflict with the Canadian Autoworkers union (CAW). The CAW sponsored a research project at CAMI that included regular surveys (administered during face-to-face interviews) of employee attitudes and has used what they have learned to develop a broader policy of opposition to lean production at other Canadian plants (Huxley, Rinehart, and Robertson, 1995). Furthermore, the CAW led a strike at CAMI in March 1993 (check date) -- the only strike ever to occur among the new unionized Japanese-affiliated plants in North America -- and successfully negotiated some collective bargaining restrictions on the use of lean production practices such as "kaizen" or continuous improvement activities. Among the former CAMI employees at Eisenach, I heard

LaSorda was initially assigned a team of experienced Opel managers who were fully expecting to implement production concepts that were familiar to them from past experience in Opel plants. Given Hughes's intention to make Eisenach a model plant for lean production, LaSorda asked for a team of "advisors" to help him educate and influence his Opel management team. He argued that there must be a "critical mass" of such advisors for them to have any real influence. LaSorda and Hughes were both mindful of the experience of the GM managers who trained intensively at NUMMI and were then sent, as solo change agents, to traditional GM plants. Even when placed in positions with sufficient authority, these managers were not able to accomplish much change. Hughes approved the hiring of enough advisors so they could be matched, one to one, with the management team at Eisenach -- as a kind of "shadow" management structure.

The advisors. The first set of advisors was brought to Eisenach in the spring of 1992, with most of the new plant facilities completed and while the Astra launch was being prepared. It included mostly young Americans and Canadians, primarily with engineering backgrounds, who has been working at GM Canada plants before being pegged for service at CAMI. They had therefore worked closely with Japanese managers from Suzuki who were helping oversee CAMI's launch, and had experienced the various stages of the plant start-up process. A smaller subset of the advisor group were Americans who had worked at other Japanese transplants in the U.S. and Canada -- NUMMI (California), Toyota Georgetown (Kentucky), Nissan Smyrna (Tennessee), and Toyota Cambridge (Ontario).

Opel managers at Eisenach. Aside from LaSorda, the initial management team assigned to Eisenach consisted primarily of German managers from relatively traditional Opel plants who had had no extensive exposure to lean production plants. However, given the new plan for Eisenach, two of these appointments were changed. The new production manager was German and a long-time Opel employee, but immediately before coming to Eisenach, he had worked at IBC Luton, a GM-Isuzu joint venture in the U.K. and was regarded as having a sophisticated understanding of lean production. The new human resources/labor relations manager was also Germany but from outside the automobile industry. His previous job had been as head of labor relations for a German printing firm known for its progressive work practices and creative, norm-breaking collective bargaining contracts with its unions. All other members of the management teams -- including managers of the individual departments (weld, paint, and assembly) as well as the head of engineering and quality came from other Opel plants, mostly in Germany.[3]

---

some speculation that the abrupt change in management when LaSorda left was one reason for CAMI's difficulties with the CAW.

3   No managers came from Zaragoza, which was seen as a very traditional plant, at this initial stage. Later, LaSorda was replaced by Eric Stevens, an American who had first

Opel engineers at headquarters. The R&D group (known as TDC) at Opel headquarters in Russelsheim, Germany oversees product development and process technology for GM Europe. (The corporate headquarters is in Zurich but contains no technical staff.) Their initial planning for Eisenach focused mostly on the opportunities for introducing new technology, particularly in the weld and paint shops. TDC also drew up the plans for the layout of facilities at Eisenach, which, with its three separate buildings, represented a substantial departure from the space-efficient, one building norm of most lean production plants. (see note #1 above)

However, when the emphasis shifted to making Eisenach a "lean" plant, TDC responded favorably. According to some Eisenach advisors, TDC engineers had carried out a number of study missions to Japan and to transplants such as NUMMI and believed that they understood how a lean production plant should be set up. They were particularly enthusiastic about the idea of implementing manufacturing "cells" with advanced robotic equipment in the welding shop, which would combine cutting edge technology with a new production concept. Working with long-time German equipment vendors such as robotic company KUKA, they began ordering the new equipment for Eisenach. But it would soon become clear that their vision of the Eisenach plant, both in terms of technology and the role of the workforce, differed substantially from that of LaSorda and his advisors.

East German workers. As mentioned above, Eisenach's workforce was initially drawn from the ranks of AWE employees. These employees, like most in East Germany's manufacturing industries, were relatively well-educated and many were well-versed in traditional craft skills -- although few had any exposure to advanced technologies. With the Wartburg seen as a near-luxury vehicle in the Eastern Europe context, the AWE employees had a somewhat elite status among manufacturing workers, although they labored in an ancient multistory building, carried out most operations manually, and faced frequent shortfalls of parts and other resource constraints. They had developed a reputation for being resourceful and hard-working.

While selection of workers for the immediate assembly of knock-down kit Vectras was primarily by seniority, the hiring process for the new Eisenach plant was much more elaborate. GM Europe hired an outside firm, headed by some former GM staffers, to run the selection process, following an approach used by some of the Japanese transplants in the U.S. Workers were given reading and math tests, placed in simulated production situations to check their manual dexterity, interviewed individually, and then organized into groups for exercises that allowed assessment of their willingness and ability to work in teams.

Nine weeks of training at the new plant followed, a mix of classroom instruction in lean production principles and on-the-job training. Only after this

---

worked at CAMI as production manager and then was named plant manager at Zaragoza.

period of training was the hiring decision made. In addition, the first three months of employment were considered a probationary period, at the end of which workers could be terminated. At the end of this lengthy process, a worker became a "permanent employee" and could then benefit from a number of advanced human resource policies, including a "no layoff" commitment, opportunities for further training, generous vacations, and the possibility of bonus pay.

The yield from the selection process was about 20% of those who submitted applications. In the end, only 1100-1200 of the nearly 10,000 AWE employees became part of the Eisenach workforce. Eisenach advisors told me that the long selection process had reportedly scared off some workers who had good technical skills but were uncomfortable with the interpersonal requirements aspects of the process. Rumors spread that one must be a "good talker" to be hired. As later rounds of hiring for the second and third shifts took place, applicants were increasingly drawn from a wider geographical area and didn't necessarily have previous manufacturing experience.

Union and works council. These dual channels for worker representation were established following the West German model. The first-shift production workers at Eisenach elected a works council president soon after the plant opened (date?) -- a man who had been a skilled technician and strong informal leader in the old AWE plant. The workers also became members of the IG Metall metalworkers union. The work council and union negotiated a collective bargaining agreement that bore many resemblances to NUMMI's contract -- with a management commitment to "no layoffs", few job classes, shop-floor teams, job rotation -- as well a few differences. Eisenach's contract included provisions for pay based partially on the mastery of skills -- something that is not done at NUMMI. It also placed strict limitations on management's ability to require overtime work, consistent with IG Metall policy elsewhere in Germany and different from NUMMI. This was one reason cited for making Eisenach a three-shift operation.

The compensation provisions in the contract were heavily affected by Eisenach's location in East Germany. IG Metall had recently completed industry-level negotiations with the association of metalworking companies for a five-year plant to reach parity between East and West Germany in wages and hours by 1996. Parity was to be achieved for the "Hessenter" or "tariff wage", which established the baseline pay at many West German companies. This establishes a floor for compensation but many larger companies pay above this level. Thus even with "Hessenter" parity, Eisenach's wages could potentially remain lower than GM Europe's West German plants after 1996, based on what percentage above the tariff wage level was negotiated between the company and the union in different locations.

Nevertheless, with pay rising close to West German standards in a relatively short period, this placed heavy pressure on Eisenach to achieve comparable or better productivity levels than West German plants. Both the union and works council indicated their support for achieving this goal, while management indicated their intention to do everything they could to avoid layoffs for "permanent" employees. Work council representatives, and particularly the president, were treated as members of the management team, with full access to competitive information and attending all key management meetings.

9.3
Replication efforts at Eisenach: the transfer of "lean production" concepts

The first indication that Eisenach would attempt to replicate production concepts from other plants can be found in the philosophy adopted by LaSorda and his first team of advisors. As explained by one advisor, "If a system or policy works well somewhere else, why not adopt it as our starting point?" While not ruling out the possibility of adapting the practice over time to fit Eisenach, he emphasized the importance of being willing to start by implementing "what works" elsewhere, replicated with great care.

He explicitly contrasted this approach with the one taken by TDC in its development of materials about the "Opel Production System", a modified version of the Toyota Production System. For this advisor, the problem with TDC's approach was that their modifications of Toyota Production System concepts were based on preconceptions of how they wouldn't "fit" at Opel and not on actual experience -- and that renaming and modifying every idea drawn from outside the company encouraged a "not invented here" mindset that was destructive to effective learning.[4]

Initially, LaSorda and his team of advisors found they were most comfortable replicating policies from CAMI, i.e. using CAMI as their primary "template" plant. But then they decided to draw explicitly on the wealth of experience of different advisors at other potential template plants (e.g. NUMMI, IBC Luton, Toyota in the U.S. and Canada, Nissan in the U.S.). Thus during the startup phase, advisors would meet to compare and contrast the different variants of lean production policies they had seen and to select one as their starting point. For

---

4   TDC's approach was characteristic of GM's corporate culture. Saturn's Committee of 99, made up of company and union representatives, had followed a "clean sheet of paper" philosophy in developing Saturn's production system over a period of two or three years. After surveying the range of production practices around the world, the Committee typically established policies that were a clear departure from traditional GM policies and often unique to Saturn. GM in North America had also seen a rivalry in the mid-1980s between two competing reinventions of Toyota Production System ideas -- the "Synchronous Manufacturing" program developed by Engineering and the "Quality Network" program developed by Personnel and Labor Relations (check this) -- that had caused delays, confusion, and redundancy when, eventually, both programs were implemented.

example, the materials management system was based closely on Toyota, while maintenance policies were based on Suzuki and the suggestion system was taken from NUMMI.

This insistence on replication of lean production principles also meant that LaSorda and his advisors were generally unwilling to follow policies established at GM's Zaragoza plant, which also built the Corsa and Astra products and, as noted above, was the "lead" plant for technical matters and parts sourcing. LaSorda felt that Zaragoza's traditional mass production plants would not be appropriate for the production system he was trying to establish at Eisenach, but this aroused some resentment on the part of Zaragoza managers.

Another indicator of the replication approach was the stance on process technology for the plant. While TDC was planning to introduce lots of new technology never before used in an Opel plant, LaSorda and his advisors wanted absolutely no new technology at startup. They believed in the value of the strategy followed by Toyota at NUMMI and Suzuki at CAMI of minimizing the potential sources of variation during startup by replicating as closely as possible the production process from another plant making the same vehicle. With little variation in product or process, the launch effort could focus on problems related to those things that **were** new -- training for the new workforce, resolving quality and delivery problems with new suppliers, and making sure all systems in the new facility (e.g. lights, heat, conveyors) were fully functional.

To the great frustration of many advisors, their objections about TDC's plans came too late in the process to have much effect. Eisenach became the first GM Europe plant to install a water-based paint process. In the body shop, a compromise was reached, with TDC agreeing to install more of a mix of new and older equipment than they had originally intended.

LaSorda and his advisors also insisted that they be the ones to work through initial problems with new process equipment rather than the technical experts at TDC. They wanted to insure that whatever could be learned about the new equipment during the debugging process should remain at the plant, rather than being captured by engineers from headquarters who were not involved in ongoing operations. This meant considerable discomfort for TDC engineers, who were accustomed to receiving lots of calls from plants having technical problems during startup.

The evolving strategy for how to use advisors was also affected by the effort at replication. Both LaSorda and his successor, Eric Stevens insisted on having enough advisors to match them, one-to-one, with individual Opel managers, while at the same time insisting on the adoption of lean production policies replicated from other plants -- all with the intention of giving Opel managers an incentive to learn from their advisors in an ongoing, "on-the-job", experiential way.

The success of this partnering varied considerably across the plant. One advisor told me, "Where you have two people with at least some Japanese experience together, it's very easy. But where you have an older manager with no experience outside Opel and a young advisor fresh from a Japanese transplant, it can be very tough." The role of advisor also had special demands that not all of those chosen for the job could fill. For the most part, the advisor had to work through their partner-manager -- they were not supposed to give orders or make decisions independently. For many of the young engineers filling these positions, that role was too frustrating and limited.

The first set of advisors was intended to stay at Eisenach for three years, but most left after two years, eager to return to "hands-on" responsibilities.[5] The second set of advisors, which overlapped with the first group for six months, was nearing the two year mark at the time of my visit and, for some, the level of frustration was palpable. Yet others found the position challenging but rewarding and were proud that they were the "carriers" of wisdom about lean production. (No Japanese advisors were present in the plant, presumably as a matter of policy.)

9.4
Cultural Contrasts

In this section, I will describe some of the cross-cultural interactions and contrasting perspectives that were evident during my time at Eisenach. My goal is to show how the "template" plants of CAMI and NUMMI had a larger influence on plant operations than the more immediate cultural influences at both national (e.g. Germany) and organizational (e.g. GM overall and specifically Opel in Europe) levels.

TDC vs. Suzuki. The first contrast is between TDC's and Suzuki's philosophy of technology. TDC designs (and hires vendors to build) massive pieces of equipment that are intended to last a lifetime and to tolerate comfortably the highest possible production demands, in terms of volume. Suzuki is known for its dedication to being the lowest-cost producer for certain market niches, and uses lots of relatively cheap and lightweight equipment. According to one advisor, "Suzuki equipment breaks down a lot, but their people know how to get

---

5   With GM Europe deciding to expand the use of advisors to all its plants in the last few years, many of the Eisenach advisors have gained a substantial boost in salary and responsibility upon leaving to take a position in another GM Europe plant. Managers and engineers from GM in the U.S. who have had some significant experience with Japanese companies or joint ventures are also much sought after as GM Europe seeks to expand the pool of available advisors. The greatest windfall is going to those who have actually worked for Toyota in the U.S. and Canada, since they are felt to have the most effective training in lean production principles. Finally, GM Europe has recently hired a "super-advisor" to oversee and coordinate its growing set of advisors -- a Japanese man who last worked for Toyota's internal production process consulting group in Japan and trained by Taiichi Ono himself!

it running again fast. I call it 'patch and play'. They manage to break down the organizational barriers that keep us from responding to problems as quickly."

One example will help illustrate this point. TDC's welding equipment (robots and fixtures) is heavy, large, and often securely fastened in a fixed position. One advisor pointed out a 5-pound weld gun that TDC had welded to a 75-pound metal stand, with the stand further welded to a metal plate that was cemented into the floor. However, this weld gun had a fixed welding position and a robot moved the part so that welds could be applied in multiple positions. Thus the weld gun didn't need to be fixed into place at all, since it was not moving. In contrast, Suzuki often places its equipment on wheels, with flexible hoses and cables hanging down from the ceiling to allow the equipment to be moved for better layout of the work area or when process changes are required.

Suzuki's approach offers the plant considerable flexibility in another way. As one advisor put it, "Suzuki is great at 'weekend kaizen'." Over the weekend, with production stopped, engineers can move equipment around, experiment with new software instructions for robots, try running a new part. Then they can move everything back to the way it was before production resumes on Monday morning. After several weeks or months, with all the bugs in a prospective process change discovered and fixed, the permanent changeover to the new method can easily take place over the weekend, with no production downtime.

Several differences in philosophy were also apparent in the TDC and lean production view of the appropriate role for the workforce in the production process. Opel's industrial engineers don't like team leaders ("non-productive labor") or workers who rotate across multiple tasks ("efficiency losses when changing tasks". They also don't like having workers "pull" subassemblies from an inventory rack when needed ("interruption in work effort") and would prefer having material handling employees replenish inventory whenever it is low. Finally, they dislike reduced buffers between work processes ("too much downtime") and having machines wait for people ("hurts capacity utilization") rather than vice-versa.

Eisenach is doing all of these things, following lean production principles, despite TDC discomfort. Team leaders help with coordination and problem-solving across different parts of the plant, as well as filling in for absent workers. Team members rotate not only for ergonomic reasons but also to learn enough about different jobs so they can better understand problems arising in their part of the production system. The "pull" system and reduction of work-in-process buffers reveals bottlenecks and other process problems and creates a strong pressure to resolve them quickly.

Problem-solving policies at Eisenach also challenge the traditional skill hierarchy with respect to technology at most German manufacturing plants. I observed a problem on the welding line in which the failure of a clamp to release caused the line to shut down and the andon board lights to come on. The production worker on the line was the first to go to the robot and try to fix it. With lots of simple visual controls on the robot's controller panel (e.g. a light showing the stuck clamp), it was easy for the production worker to diagnose the problem correctly. Although the production worker was then able to release the clamp, he was unable to reset the limit switch that had stopped the line. So, only about 45 seconds after the line had stopped, he summoned the maintenance man for his area. The maintenance man was able to reset the limit switch with a software command at the controller. In a typical German plant (and in most U.S. plants as well), the maintenance man would be the first to be called. Here the order was reversed, and the speed and smooth coordination with which the problem was solved was impressive.

On balance, then, Eisenach shows some signs of TDC's technology philosophy, particularly in terms of the choice of technology and the physical plant, but in operational terms, it closely followed the simple, portable technology / low buffers/ continuous problem-solving approach characteristic of lean production. More specifically, it was Suzuki's "bare bones" approach, filtered through CAMI-trained advisors, that became the dominant "template" for technological matters, and frequently counterpoised with TDC's tendency towards overengineering.

NUMMI vs. CAMI. While CAMI and Suzuki were a strong and visible influence on technology policies, NUMMI (and Toyota) was the template more often cited for policies on manufacturing logistics and the management of people. First of all, at GM Europe, Toyota was viewed as the true master (and the original innovator) of lean production, and NUMMI was viewed as the most convincing illustration of the transferability of lean production principles to another cultural context. Furthermore, NUMMI had become the template plant most often chosen as the destination for study missions for GM Europe managers and staff, either individually or in groups. As the most legitimized example of lean production within the corporate culture, NUMMI/ Toyota was viewed more favorably by Opel managers than CAMI/Suzuki, particularly since Suzuki's technology policies were so antithetical to TDC.

This was most clear for materials handling and logistics. GM Europe had managed to hire away three young Canadians who had worked in this area at Toyota's Cambridge, Ontario plant and urged them to replicate Toyota's materials system as closely as possible at Eisenach -- down to the fine details of visual control for kanban cards and FIFO (first-in, first-out) procedures for the inventory in parts storage. This was seen as working quite effectively within the plant, although the fact that many parts arrived via a two-day train trip from Spain, combined with the unreliability of truck deliveries given highway congestion, made it impossible for Eisenach to run a pure Just-in-Time system with its suppliers.

NUMMI was also seen as a more successful model for how to manage the workforce and the relationship with the union. By this time, the labor relations problems at CAMI were well-known at Eisenach, sharpening the perception that Suzuki was less skilled in knowing how to handle these issues, at least outside of Japan. NUMMI also had an important symbolic appeal with respect to the union because it had showed that a traditional union could be persuaded to support a lean production system. As noted above, many policies related to "kaizen" or quality improvement were explicitly drawn from NUMMI or other Toyota-influenced plants rather than CAMI/Suzuki.

Finally, Eisenach has established a liaison office that is modeled closely on NUMMI's and given its staff the responsibility for diffusing the lessons of Eisenach throughout GM Europe. NUMMI's liaison office was set up by GM soon after NUMMI opened, and for many years was widely viewed as ineffectual in its efforts to help GM learn from NUMMI. But the NUMMI liaison office is now seen by Eisenach as much improved in recent years. Furthermore, the staff in the Eisenach liaison office believe that the past failings of the NUMMI office make it instructive as a "template", as much as its more recent successes.

The first key lesson the Eisenach liaison office has drawn from the NUMMI experience is the importance of exposing "students" (i.e. managers and staff visiting from other plants) to the daily operations of the plant in an experiential format. Plant tours and classroom instruction are of limited value in conveying the production system principles that NUMMI relies upon. Instead, time spent studying a single area of the plant, even working for a time on the line, and seeing how different problem situations are handled is far more effective as a way to learn..

The second key lesson is that the exposure to NUMMI is most effective if staggered (several short visits over the course of a year, as opposed to one long stay) and if linked to a project being carried out at the home plant. At the time of my visit, Eisenach's liaison office was experimenting with several designs for its programs for managers and staff from other parts of GM Europe, all involving iterations of time at Eisenach developing a project "action plan" for their home plant, following by implementation efforts at their home plant, followed by another stint at Eisenach.

So despite the fact that the overwhelming majority of Eisenach advisors had past experience at Suzuki, Toyota-managed plants -- especially NUMMI -- were more often cited as the appropriate and legitimate template for the policies being implemented at Eisenach.

Cultural identities at national vs. company vs. local levels. It became clear during my interviews that all of the advisors and many of the senior German managers at Eisenach were "marginal" with respect to certain cultural norms associated both with Germany and with the larger GM culture. Through their previous experiences, they had come to believe that the dominant model within their country and company was not working and needed to be changed. They were drawn to Eisenach precisely because it offered an opportunity to establish different norms, implement different policies, and try out new behaviors. Many saw themselves as having been restless and uneasy at earlier stages of their career at GM, whether in Europe or in the U.S. They prided themselves on having wanting to do things differently when working within the traditional system.

The two German managers brought in after the plant's reorientation towards lean production -- the production manager with experience at a Japanese transplant in the U.K. and the personnel manager with experience in the printing industry -- were, in many ways, a better fit to Eisenach than they would have been at more traditional Opel plants.   However, they did sometimes find themselves caught in the middle of disagreements between the more traditional Opel managers and the young American and Canadian advisors. The advisors also tended to "fit" better, in a cultural sense, at Eisenach that they would have at a more traditional plant -- although some were impatient with the advisor role, as noted above, and sought more responsibility at a site that would be compatible with their production system views.

Thus the cultural lens most useful for understanding the views and behaviors of Eisenach managers and advisors was not their country of origin or their corporate affiliation, but whether they had had experience working at a Japanese-managed lean production plant (or, in the case of the personnel director, a progressive company outside the automobile industry.) To call the status of this Eisenach group "marginal" with respect to their cultural· and organizational backgrounds is misleading if it implies isolation, alienation, and estrangement. Rather, marginal status was almost a source of pride for this group, who believed they were part of the "wave of the future" for GM and, to a lesser extent, for Germany as well. However, they did express concerns about what their future career experiences within GM Europe might be like after leaving the comfortable culture created at Eisenach.

West German company vs. East German workers. Several managers and advisors told us that East German workers were, in some ways, more inclined towards the behaviors expected under lean production than most West German workers. They were comfortable and familiar with working in groups, since informal working groups were encouraged in the old AWE factory and the norms in most Eastern European countries were less individualistic and more group-oriented than in the West. Team members at Eisenach frequently met after hours to socialize, aiding team cohesiveness and blurring the line between work and non-work social relationships in a way more familiar in Japan than in most Western countries. Furthermore, workers were quite accustomed to being very resourceful on the job and finding ways to solve problems with minimal resources. At the AWE plant, workers often had to find a way to "jury rig" old machines to keep them running, or to salvage scrap materials in order to build some fixture needed on the assembly line. This custom of coping with unexpected situations in resourceful and clever ways was a good fit to the "kaizen"˳ philosophy of continuous improvement that the Eisenach plant was trying to encourage.

There were, however, a few areas in which workers were reluctant to become involved in problem-solving activities. They were often afraid of the consequences of making mistakes, so they were very cautious in the suggestions ·they would make. The Eisenach advisors also reported having difficulty persuading workers to accept responsibility when, for example, a small task force was being set up to deal with a quality problem or advance planning for a new model launch.

Furthermore, workers often had emotional reactions when problems did occur during production, feeling both angry and ashamed. When the plant was having difficulties during the launch of the Corsa, workers were anxious and complained to managers, saying "you told us how great this new production system was and how well it would work -- so why are we having so many problems?". Once when the plant lost several hours of production in the weld shop one week, due to errors made by workers still in training, the whole weld department offered to come in and work on Saturday to make up the lost production.

The fact that the East German workers had no previous experience in a modern, Western mass production setting probably made it easier for them to accept the lean production principles and policies being introduced at Eisenach. As their opportunities for comparison with other, more traditional workplaces increase, through the economic development of the former East by companies from the West, they may be more likely to challenge the unconventional approach taken at Eisenach. On the other hand, if Eisenach thrives and continues to maintain the high commitment levels of its East German workforce, it may provide an intriguing example of a move directly from older craft traditions to new lean production organizing principles without passing through an intervening stage of mass production.

Eisenach vs. the larger GM system. Eisenach's very success at creating a plant culture that differs substantially from the norms of the larger GM system also makes them vulnerable to criticism and interference. At the time of my visit, the plant managers and various advisors described a certain tension within GM Europe about whether Eisenach would be allowed to stay a "learning lab" for other plants, with appropriate support for the liaison office and for the staff time spent teaching and training outside visitors, or whether it would be forced to become more "integrated" into the company. Clearly, for the Eisenach management team, "integration" would mean a loss of independence and a reduction in resources. They believed that the only reason that some senior managers at GM Europe were pushing for more integration was jealousy about the attention and resources given to Eisenach.

Some advisors saw an analogy to Saturn, GM's independent subsidiary making small cars in the U.S. Saturn too was intended as a "learning lab" for the rest of GM but by most accounts, very little of what has been learned at Saturn, from either its successes or its difficulties, has made its way into other GM plants. Saturn was also seen as benefiting from extraordinary resources and patience during a lengthy startup, at a time when other GM divisions were starved for investment. And Saturn also faces increasing pressures to become more "integrated" with the rest of GM -- to build products developed by other divisions, to expand production at other existing GM plants rather than at its

Spring Hill location, and to accept more surplus workers and managers from the rest of GM during a period of radical downsizing.

For the Eisenach plant -- which has replicated a different production system, created a different culture, and attracted a talented group of managers and engineers who see themselves as different from the prevailing corporate norms -- there is a risk that the larger GM system will not tolerate their differences and will force them, in various ways, to conform. Thus Eisenach's future may depend on which of two influence processes is more effective -- Eisenach's efforts to teach the rest of GM Europe about lean production, or the larger GM system's efforts to prevent Eisenach from being treated as a special case.

9.5
Discussion

The Eisenach case study illustrates, I would argue, a distinctive strategy for the diffusion of new production system concepts in the automotive industry -- here called "replication". The replication process first requires good access and opportunities for study at a "template" or "learning model" site, and then a means of transferring knowledge about the template to the new location. In Eisenach's case, there were multiple template sites and a group of advisors with direct experience at those sites who acted as "carriers" of their philosophies, policies, and culture.

The effort to make Eisenach the "NUMMI of GM Europe" reflects the legitimization of the organizing principles of lean production as a new dominant model, arising from various sources (the business press, the MIT auto project, corporate benchmarking exercises) and strongly backed by the top corporate leadership of GM Europe (in sharp contrast with the leadership of GM in North America). As a result, I would argue, Eisenach looks and feels more like other lean production plants, including those that are Japanese-managed, than it looks and feels like other plants at Opel or GM in the U.S. This appears to support the argument of increasing convergence towards a new dominant model worldwide, but is also consistent with the prediction of increasing divergence within country and company due to variations in the extent and speed of the diffusion of lean production concepts.

The primary template sites evoked during replication efforts at Eisenach were CAMI and NUMMI. While a majority of the managers and advisors from North American had direct experience working at CAMI, NUMMI was the template site more strongly legitimized as the model for change within the rest of GM Europe. Furthermore, through CAMI and NUMMI has come the influence of Suzuki and Toyota corporate cultures, even though there are only a few individuals who were ever employed at either Japanese company.

It is probably only in replication situations that the culture of the template site is so strongly felt at the new site, since key managers are bringing experiential knowledge from the template site. Where a dominant model is imitated without much direct access to a template site (e.g. where Western companies decide to implement the Toyota Production System based on what has been written about it rather than through access to Toyota plants), the influence of the cultural component of the template site should be much less.

At Eisenach, country-of-origin cultural differences are still apparent in some struggles between traditional Opel managers and young American advisors. But to characterize these interactions in terms of the modal tendencies of each national culture (or even Opel vs. GM in the U.S.) may be misleading, given that these individuals tend to see themselves as marginal with respect to norms for their country or company.

This case suggests that individuals, as "carriers" of cultural norms from template sites, are only likely to have a strong influence on the culture of a new site if their numbers are large or if the cultural norms they bring are strongly legitimized both within the company and the industry. It also implies the need to pay more attention, in our theories of cross-cultural interaction, to the "invisible" cultural players -- the cultural norms drawn from a dominant model that is highly legitimized in the institutional environment -- in situations where national and company cultural influences are already at play.

9.6
References

Brannan, Mary Yoko and Jane Salk. 1994. "Cultural Dynamics in Bicultural Work Settings: Empirical Examples of Negotiated Culture," presented at Academy of International Business conference, Boston, MA, Nov. 1994.

Cusumano, Michael. 1985. The Japanese Auto Industry: Technology and Management at Toyota and Nissan. Cambridge, MA: Harvard University Press.

DiMaggio, Paul J. and Walter W. Powell. 1991. The New Institutionalism in Organizational Analysis. Chicago: University of Chicago Press.

Florida, Richard and Martin Kenney. 1991. "Transplanted Organizations: The Transfer of Japanese Industrial Organization to the U.S." American Sociological Review Vol. 56 (June), pp. 381-398.

Hofstede, Geert, M.H. Bond, and C.L. Luk. 1993. "Individual Perceptions of Organizational Cultures: A Methodological Treatise on Levels of Analysis," Organizational Studies, Vol. 14, No. 4, pp. 483-503.

Kogut, Bruce and Udo Zander. 1992. "Knowledge of the Firm, Combinative Capabilities, and the Replication of Technology." Organization Science, Vol. 3, No. 3 (August), pp. 383-397.

Kogut, Bruce. 1993. "On the importance of being inert," in Organizational Theory and the Multinational Corporation, Sumantra Ghoshal and D. Eleanor Westney (eds.). New York: St. Martin's Press.

Lewchuk, Wayne.   1988.   American Technology and the British Car Industry. Cambridge, England: Cambridge University Press.

MacDuffie, John Paul. 1995. "Human Resource Bundles and Manufacturing Performance: Organizational Logic and Flexible Production in the World Auto Industry," Industrial and Labor Relations Review, Vol. 48, No. 2 (January), pp. 192-221.

MacDuffie, John Paul and Frits Pil.   1995.   "Work Practices and Manufacturing Performance in the World Auto Industry: The Diffusion and Adaptation of Lean Production in the U.S., Europe, and Japan," submitted to Industrial Relations.

Nelson, Richard and Sidney Winter. 1982. An Evolutionary Theory of Economic Change. Cambridge, MA: Belknap Press.

Szulanski, Gabriel. 1994. "Unpacking Stickiness: An Empirical Investigation of the Barriers to Transfer of Best Practice Inside the Firm," working paper.

Tushman, Michael and Elaine Romanelli.   1985.   "Organizational evolution: A metamorphosis model of convergence and reorientation," in B.M. Staw and L.L. Cummings (eds.), Research in Organizational Behavior, vol. 7: 171-222. Greenwich, CT: JAI Press.

Weiss, Stephen. 1994. "Culturally-Responsive Strategies in Negotiation: Developing Empirical Tests and Training Implications," presented at Academy of International Business conference, Boston, MA, Nov. 1994.

Westney, Eleanor. 1987. Imitation and Innovation. Cambridge, MA: Harvard University Press.

Winter, Sidney. 1995. "Four Rs of Profitability: Rents, Resources, Routines, and Replication," in Resource-Based and Evolutionary Theories of the Firm: Towards a Synthesis, Cynthia Montgomery (ed.), Hingham, MA: Kluwer Academic Publishers, pp. 147-178.

Womack, James, Daniel Jones and Daniel Roos. 1990. The Machine That Changed the World. New York: Rawson-MacMillan.

10　The Authors of the Book

Hisanaga Amikura
is Associate Professor of Business Administration at the Faculty of Economics at
the Sophia University, Japan. He received an MBA from Graduate School of
Commerce, Hitotsubashi University. His research and publications have focused
on organizational change, product development and production system.

Arnaldo Camuffo
is an associate professor of Human Resource Management at the Department of
Business Economics and Management at the Venice University of Ca' Foscari. He
holds a Ph.D. in Management at the Venice University and a Master of Science
in management from Sloan School of Management of the Massachussets Institute
of Technology. He is a research associate at the International Motor Vehicle
Program (IMVP) of the M.I.T. and is the italian representative of the Tim
division of the Academy of Management. He has published books and arcticles
both in Italy and abroad.

Anna Comacchio
is an assistance professor of Human Resource Management at the Department of
Business Economics and Management at the Venice University of Ca' Foscari.
She holds a Ph.D. in Management at the Venice University. Her research interest
are organization and human resource management, competency management and
innovation process. On these issues she has published books, essays in books and
articles.

Kajsa Ellegard
Dr., associate professor, Department of Human and Economic Geography, School
of Economics.and Commercial Law, Gothenburg University. She works in the
tradition of time-geography and her empirical studies are related to the
automobile industry and to the everyday living conditions of individuals in
households. Her first study concerned the development of new work organization
principle in the process of automation in carbody shops whitin Volvo - followed
by an actor-oriented study during the planning of Volvo - Uddevallaverken. She
holds a research position in Time Geography at the Swedish Council for
Research in Humanities and Social Science.

Charles H. Fine

He is Associate Director of the Center for Technology, Policy and Industrial Development and teaches in the operations management group at MIT's Sloan School of Management. He holds a M.S. in Operations Research from Stanford University in 1981, and a PhD in Business Administration (Decision Sciences) from Stanford University in 1983. He also serves as Director of the Technology Supply Chain project, Co-Director of MIT's International Motor Vehicle Program and Co-Director of the Fast and Flexible Manufacturing Project. Current work addresses technology sourcing decisions and supplier relations -- primarily in the automotive and semiconductor industries, with a particular focus on manufacturing equipment development and sourcing.

Michel Freyssenet

Dr., sociologist, is a research director at the French National Research center (CNRS). He is a co-director of the Gerpisa (Permanent Group for the study of the automobile Industry and its Employees) and the International Program *The emergence of the New Industrial Models*.

Urlich Jurgens

Dr., is a senior researcher at the Social Science Research Center Berlin (WBZ) and .Privatdozent (external professor) at Berlin Free University. He has directed the internationally comparative projects in the fields of company strategies, industrial relations and work organization with a focus on the automobile industry.

John Paul MacDuffie

is an assistant professor in Management Department at the Wharton School of Business, University of Pennsylvania. He received his BA degree from Harvard University and his Ph.D. degree from the Sloan School of Management at M.I.T. (Massachusetts Innsitute of Technology). His research explores policies in manufacturing settings. With Frits Pil, he has completed the second survey for the International Assembly Plant Study, from a sample of 88 automotive assembly plants representing 20 companies and 21 countries, under the sponsorship of the International Motor Vehicle Program (IMVP) at M.I.T.

Giuseppe Volpato

is a professor of Management and at Venice University Convenuta' Foscari. He is currently involved in the ICDP within the IMVP of M.I.T. and the Gerpisa of the University of Evry Val d'Essonne. He has carried out extensive research in the automobile industry all over the world, participating and giving presentations at a number of conferences. He has published books, essays in books and articles in Italy, the UK and France.

K. Shimokawa, U. Jürgens, T. Fujimoto (Eds.)

Transforming Automobile Assembly
Experience in Automation and Work Organization

Five years after the publication of MITs lean-production book, practitioners and academics from Japan, USA and Europe present new concepts, findings and conclusions in regard to one of the most critical areas of automobile production. The focus is to explore automation and work organization for the final assembly operations in the world's automobile industry. The authors are company practitioners in charge of planning assembly operations and academic researchers drawing from recent empirical work. Thus, the book presents a multi-facetted view on a development of critical importance for future development of the industry. Rich with figures, photos and tables, the text is vivid, easy to understand and illustrative.

1997. X, 414 pp. 140 figs.
Hardcover DM 98,-
ISBN 3-540-60506-1

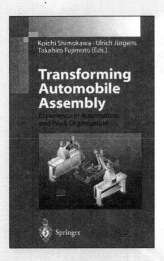

Springer
and the
environment

At Springer we firmly believe that an international science publisher has a special obligation to the environment, and our corporate policies consistently reflect this conviction.

We also expect our business partners – paper mills, printers, packaging manufacturers, etc. – to commit themselves to using materials and production processes that do not harm the environment. The paper in this book is made from low- or no-chlorine pulp and is acid free, in conformance with international standards for paper permanency.

 Springer

Printing: Mercedesdruck, Berlin
Binding: Buchbinderei Lüderitz & Bauer, Berlin